Fuel Rights Handbook

18th edition

Alan Murdie, Cecilia Torsney, Alison Gillies and Energy Action Scotland

Child Poverty Action Group works on behalf of the one in four children in the UK growing up in poverty. It doesn't have to be like this. We use our understanding of what causes poverty and the impact it has on children's lives to campaign for policies that will prevent and solve poverty – for good. We provide training, advice and information to make sure hard-up families get the financial support they need. We also carry out high profile legal work to establish and protect families' rights.

Published by Child Poverty Action Group
30 Micawber Street
London N1 7TB
Tel: 020 7837 7979
staff@cpag.org.uk
www.cpag.org.uk

A CIP record for this book is available from the British Library

ISBN: 978 1 910715 10 9

Child Poverty Action Group is a charity registered in England and Wales (registration number 294841) and in Scotland (registration number SC039339), and is a company limited by guarantee, registered in England (registration number 1993854). VAT number: 690 808117

Cover design by Colorido Studios
Typeset by David Lewis XML Associates Ltd
Printed in the UK by CPI Group (UK) Ltd, Croydon CR0 4YY

The authors

Alison Gillies is a welfare rights worker at CPAG Scotland.

Alan Murdie LL.B Barrister, is the chairman of Nucleus Legal Advice in London.

Cecilia Torsney is a debt adviser at the Mary Ward Legal Centre in London.

Energy Action Scotland is the national charity working for an end to fuel poverty and to promote warm, dry homes for all in Scotland.

Acknowledgements

The authors would like to thank everyone who has contributed to this book. In particular, thanks are due to Elizabeth Gore, Helen Melone, Apostolos Gkrimpas, Robert Kerr and Mark Willis for their invaluable comments and assistance. Thanks too to the previous authors for their contribution to the book.

We would also like to thank Nicola Johnston for editing and managing the production of the book, Anne Ketley for the index and Pauline Phillips proofreading the text.

The law covered in this book was correct on 1 October 2016 and includes regulations laid up to this date.

Contents

Contents

Abbreviations

AA	attendance allowance
CA	carer's allowance
CERT	Carbon Emissions Reduction Target
CTC	child tax credit
DBEIS	Department for Business, Energy and Industrial Strategy
DECC	Department of Energy and Climate Change
DRO	debt relief order
DLA	disability living allowance
DNO	distribution network operator
DWP	Department for Work and Pensions
ECO	Energy Company Obligation
EPC	energy performance certificate
ESA	employment and support allowance
EHU	extra help unit
EST	Energy Saving Trust
FITs	feed-in tariffs
FCA	Financial Conduct Authority
HB	housing benefit
HEEPS	Home Energy Efficiency Programmes for Scotland
IB	incapacity benefit
ICO	Information Commissioners' Office
IS	income support
JSA	jobseeker's allowance
MCS	Microgeneration Certification Scheme
PC	pension credit
PRHP	Private Rented Housing Panel
PV	photovoltaic
RHI	Renewable Heat Incentive
SLC	standard licence conditions
TCR	tariff comparison rate
TIL	tariff information label
UC	universal credit
WHD	Warm Home Discount
WTC	working tax credit

Chapter 1

Introduction

This chapter covers:
1. Sources for your rights (below)
2. The structure of the industry (p2)
3. How to use this book (p6)

1. Sources for your rights

The sources to refer to for your rights, in respect of the supply of gas and electricity, are as follows.

- Primary legislation – principally Acts of Parliament, the Gas Acts 1986 and 1995, the Electricity Act 1989, the Competition and Services (Utilities) Act 1992, the Utilities Act 2000, the Energy Act 2010, the Energy Act 2011, the Energy Act 2013 and the Consumer Rights Act 2015; the legislation particular to Wales is also passed by the Welsh government and in Scotland by the Scottish government.
- Statutory instruments – regulations made under legislation – eg, the Electricity (Standards of Performance) Regulations 2015 SI No.699 and the Gas and Electricity (Consumer Complaints Handling Standards) Regulations 2008 SI No.1898.
- Law reports/court decisions of judgments and rulings by the higher courts which clarify the scope and meanings of words and phrases used in legislation.
- Licences – the Utilities Act 2000 amended both the Gas Act 1986 and the Electricity Act 1989, changing the licensing regimes for both the gas and electricity industries.
- Your contract with your supplier – if you get your fuel from one of the licensed gas or electricity suppliers the rules governing your relationship with that supplier are in the legislation, in statutory instruments or arise from the contract. The standard terms and conditions for your contract must be freely available from the supplier.
- Codes of practice – each supplier publishes its own code of practice or statement of policy for various processes, such as complaint handling, marketing or billing and the installation of smart meters. The codes are not legally binding by themselves, but they do indicate how a supplier should and usually will behave in certain situations. You may be able to get a remedy

against a supplier's practice or particular action simply because it breaches one of the relevant codes of practice. Copies of the complaint handling code of practice should be made available to any person that requests it and should also be published on the supplier's website. Regular advisers in this field should have the relevant codes for the main suppliers in their locality; the codes may also be referred to in legal proceedings.[1]

- The gas and electricity minimum standards of performance regulations. Regulations passed by Parliament lay down minimum standards for the performance of gas and electricity supply companies and distributors for various situations.

- Decisions of the Ombudsman – although not binding, these give an indication of the standards expected and can help assess the adequacy of responses to complaints.

- Objectives for tackling fuel poverty are set by the Energy Act 2013 and a fuel poverty indicator for England was adopted by the government, following the outline of its intentions in the strategy document *Fuel Poverty: a Framework for Future Action*.[2]

- Regulations and directives issued by the European Union (EU). These provide a framework within which much of the law governing energy is constructed. The form that many aspects of energy law will take – ranging from charges on cross border supply to consumer protection – when the UK ceases to be a member of the EU following the referendum vote to leave in June 2016 is an area of great uncertainty. As five of the 'big six' energy providers are companies owned and based outside the UK, this is likely to be a major issue. However, until the withdrawal of the UK from the EU, existing provisions continue to apply, and any new regulations issued will have effect until that date.

---

2. **The structure of the industry**

Since 1999, all gas and electricity customers in Great Britain have been able to choose the company from which they buy their fuel supplies.

Gas

The gas industry is split into three parts – shippers, transporters and suppliers – with a requirement on those operating in each part to be licensed. Shippers buy gas and put it into the pipes, transporters convey it to your meter and suppliers sell the gas to you (shipping and supplying is normally done by different parts of the same company).

The main effect for you is that the supplier who sends the gas bill does not actually handle the gas itself – that is the role of the transporter. If there is a gas leak, for example, you should contact the transporter, not the supplier. The main gas transportation network is split up into four companies:

- National Grid (formerly Transco) transports gas in four areas of England: north west England, London, eastern England and the west midlands, covering approximately 11 million users;
- Northern Gas Networks cover 2.7 million users in the north of England;
- Scotia Gas Networks cover 5.9 million users in Scotland and southern England;
- Wales and West Utilities are available throughout Wales and the west of England.

In addition, there are also a number of independent gas transporters that have various smaller networks throughout Britain. All emergencies, including gas leaks, emissions of carbon monoxide and fires and explosions, are reported directly to the National Grid Gas Emergency Line on 0800 111 999, no matter who the gas transportation company is.

Electricity

Private companies involved in the generation, transmission, distribution and supply of electricity are required to be licensed. There are no longer regional monopolies.

National Grid owns the electricity transmission system in England and Wales. Local distribution is still done by one of the 14 former public electricity companies, the supply of electricity is entirely commercial and therefore (in theory) competitive, and you are able to select the supplier from whom you buy your electricity and transfer from one to another. In the majority of cases, this amounts to a right to choose between one of six major suppliers and a number of smaller licensed energy companies, unless you are able to access a smaller supply or establish some degree of independent generation of energy. Companies are obliged to make available a written statement giving summary of your rights as a customer and the expected standards of performance in law when supplying you with power.[3]

Contracts

Your gas and electricity are supplied under a contract or deemed contract from the supplier/s of your choice. 'Dual fuel' contract suppliers can supply both gas and electricity under contract.

In theory, contracts are reached by negotiation and agreement. In practice, most terms and conditions are presented to consumers on a 'take it or leave it' basis within the framework of general licensing conditions imposed upon supply companies. The use of contracts means that finding out about your rights is now far more complicated than it used to be prior to privatisation of the energy industry. Because the range of different tariffs and contracts potentially available to consumers had become so complex between 1999 and 2012, the Energy Act 2013 imposed that a supplier can now offer no more than four tariffs to a domestic customer for each fuel and meter type at any one time.[4]

You need to look particularly at the contract given to you by your supplier. There will be important differences in the terms of the contract when compared with those of other suppliers.

Ofgem: the industry regulator

Ofgem was set up in March 2000 to replace the separate regulatory bodies for the gas and electricity industries. The main functions of Ofgem are promoting competition in all parts of the gas and electricity industries and regulating them.

Ofgem's regulatory functions include granting licences, monitoring performance, regulating the areas where competition is not so effective (such as the monopoly on pipes and wires) and determining the strategy for the fuel industry.

Ofgem has the power to fine energy companies up to 10 per cent of their annual turnover for regulatory breaches. Those requirements are principally set out in the Electricity Act 1989, Gas Act 1986 and regulated company licenses, and include rules on sales practices and complaint handling.

Following the Energy Act 2013, Ofgem has been given wider powers to obtain consumer redress orders. These build upon existing powers to impose penalties and allow it to direct compensation to consumers who have been adversely affected by breaches of licence conditions by suppliers.[5] From 2016, Ofgem has also been granted a power to act as an enforcement body for certain breaches of consumer protection.

Consumer protection

Since privatisation, a series of official bodies have been responsible for consumer protection including Consumer Focus, Energywatch and the National Consumer Council. From April 2014 this role has been taken on by Citizens Advice and Citizens Advice Scotland. In addition to providing dedicated telephone lines providing advice on consumer protection, these bodies also run campaigns to promote awareness of energy saving measures and means of reducing bills.

Unfair terms in supply contracts can be made subject to enforcement activity by a range of authorities including the Competition and Markets Authority, Ofgem and local authorities' trading standards departments (see Chapter 14).

Companies can voluntarily choose to compensate consumers who lose out as a result of their wrongdoing. The Energy Ombudsman can also force the firms to pay consumers up to £10,000 if it deems complaints about 'energy bills, sales activities, problems arising from switching supplier or with the supply of gas and electricity' to be legitimate.

Consumers also have the right to bring their own private legal actions as individuals. You may also seek redress for some aspects of wrongdoing, such as breach of contract, through the civil courts.

Minimum standards of performance

Minimum standards of performance for energy suppliers and distributors are set out in regulations. The Electricity (Standards of Performance) Regulations 2015, the Gas (Standards of Performance) Regulations 2005 and the Electricity and Gas (Standards of Performance) (Suppliers) Regulations 2015 set out minimum standards of service for consumers. If a supplier or distributor fails to meet these standards, compensation is payable to the consumer as set down in the regulations. Section 13 of the Supply of Goods and Services Act 1982 as amended by the Consumer Rights Act 2015[6] provides that a term requiring that a service to a consumer is undertaken with reasonable competence and skill must be included in every consumer contract.

Energy efficiency and fuel poverty

The price that households pay for their energy is largely driven by movements in fossil fuel prices. However, successive governments have sought to mitigate the effects of such prices by establishing a range of policies aimed at helping consumers obtain a better deal from the market. Over the last decade, greater emphasis has been placed on the role of energy efficiency in achieving a number of strategic aims, including the establishment of a sustainable energy policy for the UK, with policy goals being pursued to 2030 and in some cases to 2050.

However, the Energy Act 2013 recognised that a wider range of measures and mechanisms were needed to help consumers. The Department of Energy and Climate Change (DECC) published an Annual Energy Statement in 2013 and 2014 setting out policies and goals and progress made. In the October 2013 statement, the government identified a key objective of helping households take control of their energy bills and keeping their costs down. This has been accompanied by a number of market reforms aimed at simplifying tariffs, encouraging consumer switching and moving vulnerable consumers on to cheaper tariffs to enable households make significant energy bill savings. However, the government has ruled out intervention by the Secretary of State to regulate prices or any introduction of energy price controls such as those which operated in the mid-1970s.[7]

The government believes that supplying consumers with accurate information about their consumption and use of energy is a key component of reducing energy bills. Since 2011, it has been promoting a programme of replacing existing meters with 53 million smart meters under a central change programme managed in its first phase by Ofgem and the second by the DECC (now the Department for Business, Energy and Industrial Strategy).[8] Most households will have smart meters installed by their energy company between 2015 and 2020. This has occurred along with efforts to establish a co-ordinated network of third sector organisations that work with advisers to support vulnerable households to engage with the energy market.[9] Ofgem continues to provide regulatory functions in respect of the programme, including the regulation of the Data and

Communications Company (DCC) which is responsible for linking domestic smart meters with the systems of energy suppliers, network operators and energy service companies. DCC will develop and deliver the data and communications service through external providers. A code of practice for the installation of smart meters has been issued.[10]

Also following on from the Energy Act 2013, the government amended the Warm Homes and Energy Conservation Act 2000 to put in place a legal framework requiring a new fuel poverty target for England. This led to the issue of the Fuel Poverty (England) Regulations 2014 which aim to ensure that people living in fuel poverty in England have homes with a minimum energy efficiency rating of Band C as determined by the Fuel Poverty Energy Efficiency Rating Methodology (dated 17 July 2014).[11] The government has set a target of 31 December 2030 for achieving this objective.[12] For England, a Parliamentary Committee on Fuel Poverty has been established to monitor progress and conduct research and reports – the next report is due in 2017.[13]

Fuel poverty is largely a devolved matter. The provisions of the Warm Homes and Energy Conservation Act 2000, as they apply to Wales, led to the Welsh government's Nest scheme which provides a range of free, impartial advice and support to reduce energy bills and, for those who are eligible, a package of free home energy efficiency improvements (see p173). Scotland has its own fuel poverty definition/indicator and legislation relating to fuel poverty with a target to eradicate fuel poverty by November 2016. Although improvements in energy efficiency in homes were recorded, this outcome was not realised.[14] Also, an increased mortality rate for winter deaths in 2015/16 at 2,580 was reported by Energy Action Scotland.[15]

3. **How to use this book**

Unless specified, everything in this book applies to both gas and electricity. The main legislation applies to Great Britain (England, Wales and Scotland) only. Northern Ireland is, therefore, not covered. Where the law in Scotland differs, this is noted.

Use this book principally for help in tackling fuel poverty – ie, the inability to afford adequate warmth in the home. That has always been this *Handbook's* main purpose. It does not aim to cover policy issues or examine background information in detail, but to act as a guide to the rights of consumers and the actual problems that consumers face in practice. Other publications and organisations which may be able to help are listed in Appendix 1.

This *Handbook* consists of two parts:
- chapters dealing with various topics. Look at the contents at the beginning of each chapter and consult the index to find the topic you are seeking. References at the end of each chapter give the sources of information so that you can use

them as an authority for actions; chapters 13 and 14 also give a number of key legal sources where further information can be found;

- appendices, which contain supplementary material and information.

There are references in the text to other CPAG handbooks which provide more detail on specific topics, such as benefits and dealing with debt. Where detailed information is required, such as eligibility criteria for benefits, consult the specialist handbook.

Abbreviations are used in the text to save space. The abbreviated term is explained in full the first time it is used in a section, and on pviii there is a list of all the abbreviations used.

The references in the text and notes to Standard Licence Conditions refer to the versions which were published on 13 July 2016 for electricity and 10 August 2016 for gas (available on the Ofgem website). Generally, the numbering for gas and electricity is the same, but where it differs, both numbers are shown.

Notes

1. Sources for your rights
1 *Laverty and others v British Gas Trading* [2014] EWHC Ch 2721
2 DECC, *Fuel Poverty: a Framework for Future Action*, Cm 8673, July 2013
3 Reg 22 E(SP) Regs; reg 10 EG(SP)S Regs

2. The structure of the industry
4 s139 EA 2013
5 Report Stage of the Energy Bill 2013 *Hansard* 5 February 2013 ; s144 EA 2013
6 s60 and Sch 1 paras 37 and 38 (c) CRA 2015
7 Energy Bill debate, Public Bill Committee, 7 February 2013, *Hansard* col 528
8 DECC, *Update Statement,* Autumn 2015
9 DECC, *Ensuring a Better Deal for Energy Consumers: discussion document,* November 2012
10 *Smart Meter Installation Code of Practice* (SMICoP), 27 June 2016, available at www.smicop.co.uk
11 Reg 2(2) FP(E) Regs
12 Reg 2(3) FP(E) Regs

13 www.gov.uk/government/ organisations/committee-on-fuel-poverty
14 Scottish Government, 'Fuel Poverty Strategic Working Group', 29 June 2016
15 www.eas.org.uk/en/increased-winter-mortality-excess-winter-deaths_50538/

Chapter 2

· ·

Choosing a supplier

This chapter covers:
1. Suppliers and switching (below)
2. Marketing and sales (p12)
3. Contracts (p15)

---

1. Suppliers and switching

Since 1999, consumers have been able to choose their supplier for gas and electricity and over the past decade over 50 per cent of consumers have changed their energy supplier at least once. Some customers now choose to take both gas and electricity from the same company – this is known as 'dual-fuel supply' (see p17). Currently, there are six major suppliers (British Gas/Scottish Gas, Npower, ScottishPower, SSE, EDF and E.ON) – also known as 'the big six' – who provide approximately 92 per cent of consumer supply. The remaining 8 per cent consists of around 40 smaller suppliers, such as Ecotricity, OVO Energy, First Utility and Good Energy. There has been a big increase in the number of independent suppliers in the past few years. Scotland has had a single wholesale market for electricity since April 2005.

Switching suppliers

In theory, switching supplier is fairly simple, although the process usually takes around three weeks, after a 14-day cooling-off period when you can change your mind about switching. Suppliers are responsible for managing the switch and all have signed up to standards of conduct (see p12) to ensure that customers receive fair treatment.

· ·

How do you change supplier?

1. Gather information about your current tariff, payment method and usage over the last year – you can find this on your fuel bill or on your annual statement from your supplier. Use this information to compare suppliers (see p11). If you have a smartphone, you may be able to scan the QR code on your bill. This contains all the information you need to compare and switch supplier. See Appendix 3 for examples of a fuel bill and an annual statement.

2. When you have found the best deal for you, agree a contract with a new supplier. The new supplier will write to you within seven working days to confirm the details. The new supplier will contact your current supplier for you.

See the sections below on price, comparing prices and other issues to consider before making a decision.

3. The new supplier will request a meter reading from you so that your old supplier can issue your final bill and your new supplier has the correct figure for your new bill. The new supplier will inform you of the date when your supply will be switched.

4. Check your final bill from your old supplier.

On your gas or electricity bill there is a gas meter point reference number (known as an 'M number') or an electricity supply number which is unique to your address. Once you have signed a contract which bears this number with the new supplier, the switch can take place – this should be sorted out between the new and old suppliers, although you can normally help by providing the M number or the supply number. Your present supplier may object to the transfer if you are in debt but you are still entitled to switch where the debt is below £500 *and* you have a prepayment meter (see p105). Above a £100 debt, a company has discretion whether to take you as a customer, and may do so if you have previously been a customer with a good credit history. The supplier may object where any debt is older than 28 days. In such cases, it may be difficult to switch, although the supplier should at least allow you to move on to the most favourable tariff available for the area in which you live.

It is also possible to stay with your current supplier and switch to a different tariff which is better for you.

Ofgem has rules to make the switching process 'simpler, clearer and fairer'. Following the Energy Act 2013, the number of tariffs was restricted to allow flexibility in changing tariffs[1] and Ofgem issued Standard Licence Conditions restricting what may be offered by suppliers. The number of tariffs available in any region is currently limited to four[2] and each must be only known by one name at any one time.[3] These rules aim to make it easier to compare the market. However, it is not possible to compare all available tariffs, as some are unavailable online, such as dynamic teleswitching or various time of use tariffs. Many of the switching websites say they can compare all tariffs but it is important to remember that there are more out there. Ofgem has a dedicated website (www.goenergyshopping.co.uk) to help customers look at the different options when deciding whether or not to switch and published a guide, *How to switch energy supplier*.[4]

Future changes
Ofgem referred the retail energy market to the Competition and Markets Authority (CMA) in 2014 over concerns that the energy market was not working effectively for consumers.

The CMA's final report, published in summer 2016, recommends withdrawal of certain parts of the simpler choices component of the retail market rules. Further details and any rulings by the government or Ofgem are expected by December 2016.

The following sections look at what to consider when deciding whether or not to switch and which supplier to choose.

Price

To most people, the price of their supply of gas and electricity is the most important factor in deciding which supplier to use.

See Chapter 4 for the types of meters and payment methods available.

When considering what the various suppliers are offering, look at the following.

- **Standing and unit charges** – suppliers are currently only allowed to have one structure for tariffs – a unit rate (or unit rates for time of use tariffs) and a standing charge (which can be zero). Complex tiered tariffs are banned. A 'standing charge' is a fixed monthly/daily amount that you pay the supplier for maintenance and other costs, such as maintaining connection to the power network. 'Unit charges' are the monetary amount chargeable for each unit of electricity or gas consumed and may vary considerably between suppliers, methods of payment or the type of product.
- **Payment methods** – be careful when looking at figures provided by suppliers themselves. Some advertised savings are calculated not only on the basis that you switch supplier but also that you change to a different method of payment – eg, from quarterly cash payments to direct debit. You might be able to get the same benefit with your existing supplier by switching to a different payment method. Any difference in charges for different payments must genuinely reflect the cost of the differences to the supplier.[5]
- **Penalty on default** – suppliers have always had the power to penalise customers who do not pay their bills by disconnecting them (see Chapter 8), but at least one supplier has tried introducing extra penalties. If you have difficulty with meeting all your bills on time, avoid such terms if possible. See the section on unfair terms on p19.
- **Supplier flexibility** – you might want to change your method of payment or some other aspect of your supply. For example, if you are on a prepayment meter, your current supplier might not allow you to change to a quarterly credit meter, whereas a new supplier might be more flexible. Ask different suppliers for this information.

Some suppliers offer dual-fuel supply deals (see p17) that are only available if you take both gas and electricity from them. Consider whether this would be the best for you. In particular, the convenience of a single supplier might outweigh any price disadvantages for some people.

When deciding whether or not to switch to a new fuel supplier and, if so, which one, it is best to have all the information on prices, terms and conditions so you can compare them and find the deal that best suits you. Note that suppliers are not allowed to enter into contracts through agents who require advance payments – you do not need an agent to get you a new contract and you should not use one. Full lists of all electricity and gas suppliers are available free from Ofgem. All suppliers must publish their standard terms and conditions.

Comparing prices

To make a meaningful price comparison, you need to collect information about your current supplier, payment method, annual usage and bills for the last 12 months, and then use the 'ready reckoner' comparison tables which provide a broad overview. A number of services exist allowing you to compare the price.

The potential savings available to you depend on where you live (as prices vary in different parts of the country), whether you want to switch gas or electricity supplier or both, the payment method and whether you have time of use or 'off peak' tariffs. Not all suppliers operate in all parts of the country. For details of payment methods, see Chapter 4.

There are many websites that offer to compare gas and electricity prices for you. Check that any website you use has the Ofgem Confidence Code logo. The Confidence Code sets out the minimum requirements that a provider of an internet domestic gas and electricity price comparison service (service provider) must meet in order to be, and remain, accredited by Ofgem. The prices quoted for energy deals on the accredited websites, and the information given about the offers, are shown in a fair and unbiased way. For an up-to-date list, check Ofgem's Go Energy Shopping website (www.goenergyshopping.co.uk).

The Confidence Code requires the service provider to be independent of any gas or electricity supplier. It must be a company which runs its own website and uses its own tariff database and calculating system, not merely hosting those of another service provider, and must list at least five of the cheapest suppliers. The service provider may take commission from energy suppliers, but this must not influence the information given.

Other issues to consider before switching supplier

Although the price offered by a new supplier may suit you, the other terms and conditions might not and you should check them carefully (see p18 for other terms and conditions to consider).

Before changing supplier, consider the performance and complaint handling record of the new company.

After a campaign by Which?, energy companies have to publish regular, detailed complaints data in a common format so that companies can be compared. The major energy providers have published this information on their websites quarterly and the smaller energy companies have been making this information available too.

You may be entitled to a Warm Home Discount (see p170). Before switching, check that your new supplier offers the discount and you are eligible under its criteria.

Changing supplier with existing debts

If you have owed any money for less than 28 days (eg, you have not yet paid a recent bill), you are normally able to change supplier in the usual way, and the debt will be transferred to the new supplier.

Changing supplier while in debt is subject to provisions contained in a protocol agreed between Ofgem and energy suppliers. Under these arrangements, if you have a prepayment meter you should be able to switch supplier and transfer a debt of up to £500.[6]

If you do not have a prepayment meter and you have a debt, your supplier can stop you from switching to a new supplier until you pay off your debt – this is sometimes called 'debt-blocking'. If your supplier blocks your request to switch, it must give you advice on the best tariff for you, managing your debt and energy efficiency.

You may be able to get help to clear energy debts by using funds and grants (see p181).

2. **Marketing and sales**

Marketing standards

Standard Licence Condition (SLC) 25C of the gas and electricity supply licences regulates face-to-face and telephone marketing and sales activities of licensed suppliers and their representatives. If a supplier fails to meet SLC 25C, Ofgem can take action (see Chapter 14).

SLC 25C specifically sets out rigorous standards of conduct requiring suppliers (and their representatives) to take all reasonable steps to ensure they treat consumers fairly. The standards of conduct cover three broad areas.

- **Behaviour:** suppliers must behave and carry out any actions in a fair, honest, transparent, appropriate and professional manner.
- **Information:** suppliers must provide information (whether in writing or orally) which is:
  - complete, accurate and not misleading (in terms of the information provided or omitted);
  - communicated in plain and intelligible language;

- related to products or services that are appropriate to the customer to whom it is directed; *and*
- fair both in terms of its content and in terms of how it is presented (with more important information being given appropriate prominence).
- **Process:** the supplier must:
  - make it easy for you to contact it;
  - act promptly and courteously to put things right when it makes a mistake; *and*
  - ensure that customer service arrangements and processes are complete, thorough, fit for purpose and transparent.

Information on charges

The supplier or its representative must provide you with an estimate of the annual charges for the supply of fuel under the offered contract. The details must be in writing or by way of an electronic display. A written copy must also be supplied for your own records if you subsequently enter the contract.

Compensation for mis-selling

Compensation may be payable in some cases of mis-selling; however, there is no statutory requirement within the current SLCs for suppliers to award compensation for mis-selling and Ofgem does not currently have the same statutory powers as other regulators (eg, the Financial Conduct Authority), to seek redress on your behalf. However, Ofgem does protect consumers by monitoring the energy market and taking action where there is evidence that companies have breached their obligations to consumers. Large penalties have been issued and, where possible, redress has been achieved for consumers.

The Energy Act 2013 makes provision for consumer redress orders which may be used in the future to provide an alternative to lengthy and expensive litigation.[7] The government has agreed to consider the possibility of giving Ofgem stronger and more clearly defined powers of redress in mis-selling cases in the future.

The safest way to switch without the risk of mis-selling is to use an Ofgem-accredited switching service (see p11).

See Chapter 14 for the remedies available if any of your rights have been breached. If you want to complain about a possible breach of a licence condition or a code of practice, complain first to the supplier, preferably in writing. If you need to, you can then take it further with Citizens Advice consumer service or your local authority trading standards department.

Cancelling a contract

The normal rule in contract law is that if you sign a contract you are bound by the terms and conditions in the contract, even if you have not read it.[8] However, extra protection is given to people who sign contracts at home in response to a visit by a sales representative. If you sign a contract for the supply of gas and/or electricity after an unsolicited visit from a supplier's representative, that supplier must give you a 'cooling-off period' of at least seven days.[9] This means you can cancel any contract up to seven days after you signed it (or longer if the supplier says so). Ask the representative how long the cooling-off period is if s/he does not mention it. If you cancel a contract within the relevant period you must be given back any money already paid.[10] Also, at the time you sign the contract, the sales representative must give you written notice of your right to cancel. If s/he fails to do so, the contract is not enforceable against you, whether you cancel within seven days or not.[11]

If a contract for the supply of electricity also includes providing goods or services (eg, energy efficiency measures), the charges for each must be separately identified.[12]

Misrepresentation

A signed contract may be set aside for misrepresentation at common law. This means that where you are led to enter into a contract because of a false statement of fact (not opinion or law) and the statement is untrue, you are entitled to have the contract set aside (the legal term for this is 'rescinded'). A misstatement of fact may be deliberate, negligent or innocent, but if the statement is untrue and induces you to enter a contract, then a remedy will exist in law. A court can order that a contract is rescinded and may also award damages where there has been financial loss. In practice, a supplier may be prepared to cancel a contract if there has been a misrepresentation rather than face legal proceedings. In most cases, the sums involved are £10,000 or below, the level for the small claims court procedure (see Chapter 14).

A similar rule applies to written contracts where the terms of the contract you are presented with by the seller are false or misleading. If a signature is obtained from you because the nature or contents of the agreement are wrongly described, then it is not considered binding.

Regulations also ban traders in all sectors from using unfair commercial practices towards you that prevent you from making free and properly informed buying decisions. The Consumer Protection from Unfair Trading Regulations 2008 apply to sales of electricity and gas.[13] These operate to prohibit misleading actions whereby traders supply false information or omit to provide certain information. False information includes statements which are untrue or where the overall presentation of information in any way deceives or is likely to deceive the average consumer.

Misleading information covers:
- the existence or nature of the product;
- the main characteristics of the product such as benefits and risks, availability, after-sales services;
- the extent of the trader's commitments;

- the nature of the sales process;
- the price or the manner in which the price is calculated;
- the existence of a specific price advantage;
- the need for a service, part, replacement or repair;
- your rights or the risks you may face.

Liability can also attach where the commercial practice omits material information, hides information or provides information in a manner which is unclear, unintelligible, ambiguous or untimely, or the commercial practice fails to identify its commercial intent.

The test is whether as a result of the misleading information or omission the commercial practice 'is likely to cause the average consumer to take a transactional decision s/he would not have taken otherwise' – eg, you are misled into making a decision to purchase the product or service.[14]

Offences under the regulations may be prosecuted by trading standards departments and liability may attach to both energy suppliers and subsidiary companies that act on their behalf when selling energy products.[15]

Forging of signatures

Forging of signatures is a criminal offence,[16] and compensation is payable in a case where forgery can be shown. The police or trading standards could take action in a case of forgery.

3. **Contracts**

Electricity

Supply contracts

A **'supply contract'** is an agreement for the supply of electricity to domestic premises. A supplier must not supply electricity to such premises except under a supply contract. Electricity suppliers' licences place conditions on what they are allowed to put in supply contracts (eg, terms regarding security deposits) – these are dealt with where appropriate throughout this book. This means that, when offering you a contract and supplying you with electricity, a supplier must conform to its licence conditions or face action from Ofgem (see Chapter 14).

Supply contracts are governed by Standard Licence Conditions (SLCs) 22 and 23. Contracts must be in a standard form, although there can be different forms for different areas, cases and circumstances. They must set out all the terms and conditions on which the supplier relies. If the contract is for goods and services as well as the supply of electricity, the charges for each must be separately identified. The cost of any credit element must also be shown separately.

Copies of each kind of supply contract used by a supplier must be published in a manner to secure adequate publicity. Copies must be sent to Ofgem and be available on request.[17] The contract should be provided 'within a reasonable period of time after receiving the request'. You should also be able to get information from the supplier summarising the terms of its supply contracts, with details of anything likely to influence you when deciding whether or not to take up a contract. Such information must be adequately publicised and the supplier must provide copies to Ofgem.

Gas

Supply contracts

Gas suppliers supply domestic customers with gas under the terms of a contract or a 'deemed contract' (see p17).[18] Suppliers must have a 'scheme' setting out the principal terms of contracts. The principal terms include details of the prices to be charged for gas and state if there may be any fluctuation in the amount of the bill due to variations in the amount charged by transporters to suppliers for transporting gas to your premises.

Details of the principal terms must be published in a way which is likely to bring them to the attention of the customers concerned. Ofgem must be kept informed of the suppliers' principal terms and of any variation. You are entitled to a copy of the principal terms on request and the supplier must send one within a reasonable time of receiving a request.[19] Supply contracts must be in writing.[20]

The terms of contracts may vary between different types of customer and between different areas, but not so that there is undue preference or discrimination between customers.[21] Any difference in terms and conditions offered to customers on different payment methods must reflect actual cost differences.[22] However, individual suppliers increasingly set prices on a national basis with less local variation than with electricity. Ofgem and the industry talk openly about the need for **'cost-reflective pricing'**. This means that groups of customers who are cheaper to supply may be offered discounts on the rates offered to other groups of customers.

A contract may be for an indefinite period of time, known as a 'rolling' contract', or for a fixed term. Where a contract is due to come to an end, the supplier must offer you a new contract and inform you of the terms of the 'deemed contract' (see p17) that would apply if no new contract is agreed. A supplier may not enter into a contract with you if another person has a contract with a different supplier for the supply of gas to the same premises,[23] unless that contract will have expired, or have been breached or have been terminated, before you require a supply.

Contracts for other services

Gas suppliers can offer contracts for the supply of gas together with other services – eg, service pipes or energy efficiency goods or services. Such contracts must clearly and separately identify the charges made for the supply of gas and the other services. These contracts may have different terms and conditions to contracts offered under published 'principal terms'.[24]

Deemed contracts for gas and electricity supply

A **'deemed contract'** is a contract where a customer takes a supply of electricity or gas or both in a manner otherwise than under a contract that has been expressly entered into with a supplier. A deemed contract may arise where a contract has ended without being formally renewed or where there are new occupiers who do not formally arrange a new supply contract. Deemed contracts for electricity and gas are governed by Schedule 4 paragraph 3 of the Utilities Act 2000, Schedule 6 of the Electricity Act 1989, Schedule 2B of the Gas Act 1986 and SLCs 7 and 23, and apply to situations where the supply of electricity and gas continues but the original contract is no longer in force. In such cases, you remain under an obligation to pay and the supplier is expected to behave reasonably with respect to terms and conditions and charges. Suppliers are also under a duty to use 'reasonable endeavours' to inform you of the terms and act reasonably towards you, and must not impose onerous terms.

A deemed contract continues until such time as a new contract is agreed between you and the supplier or you end the contract by leaving the premises. Schedule 6 paragraph 3(1) of the Electricity Act 1989 and Schedule 4 paragraph 3 of the Utilities Act 2000 also state:

> Where an electricity supplier supplies electricity to any premises otherwise than in pursuance of a contract, the supplier shall be deemed to have contracted with the occupier (or the owner if the premises are unoccupied) for the supply of electricity as from the time when he began to supply electricity.

Dead tariffs

Until recently, deemed contracts were traditionally higher rates. By June 2014, any consumers who were on old, expensive evergreen tariffs that are no longer open to new customers (so-called 'dead tariffs') should have been switched to their suppliers' cheapest variable rate.[25] If the dead tariff is still cheaper than the standard tariff, you are not automatically moved off it. However, suppliers are unlikely to retain customers on older cheaper tariffs.

Energy suppliers have to annually check that customers remaining on dead tariffs are not paying more than their cheapest variable rate.[26]

Dual-fuel supply contracts

Some suppliers offer both gas and electricity – this is called dual-fuel supply. There are obvious potential advantages of convenience for you if you take a dual supply.

Sometimes there may also be discounts for taking a dual supply. But check whether you will actually get a discount or other advantages – eg, you may have to deal with separate arms of the same company for your gas and electricity, which might feel little different from being supplied by two different companies. Apart from convenience, there are no other automatic benefits of having a dual supply, and you should check the terms of your contract in the same way as for any other fuel contract. Of course, dual-fuel supply contracts have to comply with both sets of provisions for gas and electricity.

If you are paying one company a single amount for both your gas and electricity, be aware of how your payment is treated. Normally, the charges for gas and electricity should be separately specified in a bill. If you make only one payment towards the cost of both gas and electricity and do not specify which fuel you are paying for, the supplier can decide which to put the payment towards. For example, if you owe £10 for gas and another £10 for electricity and then pay £10 to the supplier, the supplier can choose whether to put this towards paying off your gas bill or your electricity bill. On the other hand, if you say clearly before you pay that you are paying the £10 specifically for, say, gas, the supplier is normally bound by your decision. If you have a dispute over this, contact your supplier in the first instance, or contact the Citizens Advice consumer service. Discounts that are available on dual fuel supply contracts should normally be applied consistently on a daily basis in clear monetary terms and not as a percentage.[27]

Contractual terms

- **Period.** A contract is either for a fixed term of weeks, months or years or it is indefinite. The latter is known as a 'rolling contract'. All contracts can be terminated on 28 days' notice, but there may be a financial penalty if you terminate a fixed-term contract early. Check with the supplier whether there would be a penalty for early termination and how much it would be. In some cases, a supplier may be prepared to exercise its discretion and drop a penalty which might otherwise be imposed for early termination. Early termination is covered by SLC 24. If you want to terminate your contract after receiving notice of a unilateral change to your contract terms, such as a price increase, and switch to another supplier, your old supplier must terminate your contract within 15 days of receiving confirmation that you now have a contract with your new supplier. For more information on this process, see Chapter 5.
- **Special services.** Some contracts may be offered together with other services, such as improving the energy efficiency of your home. The costs of the supply and the services should be listed separately, including any credit element, so you can compare prices with other suppliers.
- **New products.** Some electricity suppliers offer special products. For example, for a premium (eg, a 10 per cent higher charge) some suppliers guarantee to buy enough electricity from environmentally renewable sources or from coal-

generated sources to supply your needs. Other suppliers may offer a variety of combined deals. You need to do some careful research and detailed calculations to ensure that a decision to switch supplier is based on your own circumstances, usage, location and payment method.

Unfair terms

Compared with most other businesses, gas and electricity are heavily regulated (see Chapter 14). Therefore, there should be less chance of contracts containing unfair terms and if you come across a term which might be unfair, you can complain, initially to the Citizens Advice consumer service.

However, regulation is not a guarantee. You may still need to assert your rights against unfair terms. Even better, you can try to avoid unfair terms by checking over a contract before you sign it. All terms are approved by the Department for Business, Energy and Industrial Strategy (formerly the Department of Energy and Climate Change) and reviewed by Ofgem. The scope or power of consumers to negotiate different terms (by objecting to specific terms or supplying counter terms and conditions) has yet to be tested in law in the context of consumer fuel contracts.

An **'unfair term'** is one which causes a significant imbalance in the parties' rights and obligations under the contract to the detriment of the consumer – ie, if a contractual term goes too far in favour of the supplier, it is unfair. An unfair term is not binding on you.[28] Unfair terms are governed by the Consumer Rights Act 2015 (see p227).

Contractual terms must be written in plain, intelligible language.[29] If there is any doubt as to the meaning of a particular term, the interpretation most favourable to you should be used.[30]

The courts consider that a doctrine of good faith applies in examining any term and whether it is unfair. In *Director General of Fair Trading v First National Bank plc* [2001], Lord Bingham stated:[31]

> The requirement of good faith in this context is one of fair and open dealing. Openness requires that the terms should be expressed fully, clearly and legibly, containing no concealed pitfalls or traps. Appropriate prominence should be given to terms which might operate disadvantageously to the customer. Fair dealing requires that a supplier should not, whether deliberately or unconsciously, take advantage of the consumer's necessity, indigence, lack of experience, unfamiliarity with the subject matter of the contract, weak bargaining position…It looks to good standards of commercial morality and practice.

If you come across possibly unfair terms or unintelligible language, you can refer the contract to the Competition and Markets Authority or Ofgem (see Chapter 14). The civil courts can also grant a remedy known as a 'declaration' to establish the legal effect or meaning of a term. Compensation, including for inconvenience and distress, may be awarded.[32]

Terminating a contract

Some gas and electricity contracts can be terminated on 28 days' notice. If you feel you made a mistake in changing to a particular supplier, you can give 28 days' notice and either go back to your original supplier or sign a contract with a new one. However, always check your contract – if it was for a fixed period, there may be a penalty for early termination. Some suppliers also offer long-term contracts which make provision for a 'reasonable' termination payment. Changes in the SLCs allow companies to offer fixed-term, long-term contracts without the need for extra services.

Penalties for terminating a contract

In some cases, a supplier demands payment of a fee or penalty for early termination of a contract, but a number of restrictions are placed on any power to impose such penalties.

SLC 24 for electricity provides that a termination fee shall not be demanded in the case of a contract of indefinite length (ie, a rolling contract, not a fixed-term contract) or where you have notified the supplier of an intention to terminate where the supplier has unilaterally changed or intends to change the contract. Other situations where a supplier may not impose a fee or penalty include where a property is sold or you move out or where a contract of supply is for more than 12 months or is for an initial fixed-term period.

A supplier may be prepared to waive a penalty in certain circumstances at its discretion – eg, where a contract has to be ended because you have gone into care. The supplier may also accept a lesser sum in full and final settlement of any claim for the penalty as a way of settling legal proceedings (see p100).

There may also be an argument that a supplier is under a duty to mitigate its loss (ie, take steps to reduce any loss) from early termination of the contract. The duty to mitigate is imposed at common law in a case of breach of contract. A supplier cannot just demand any sum in compensation or damages it sees fit simply because you have broken the contract in some way. The duty to mitigate losses should be referred to in correspondence to settle such a dispute.

Notes

1. Suppliers and switching
1 s139 EA 2013
2 Condition 22B2(b) SLC
3 Condition 22B2(c) SLC
4 www.ofgem.gov.uk/ofgem-publications/89430/ofg538webhowtoleaflet2.pdf
5 Condition 27.2A SLC
6 Ofgem debt assignment protocol for prepayment meter customers letter, 12 May 2015, www.ofgem.gov.uk/publications-and-updates/decision-make-modifications-gas-and-electricity-supply-licences-reform-switching-process-indebted-prepayment-meter-customers-debt-assignment-protocol

2. Marketing and sales
7 s144 and Sch 14 EA 2013
8 See *L'Estrange v Graucob* [1934] 2 KB 394
9 CP(CCCBP) Regs
10 Reg 5 CP(CCCBP) Regs
11 Reg 4 CP(CCCBP) Regs
12 Conditions 22 and 25 SLC
13 Reg 2 CPUT Regs
14 Regs 5(2)(b) and 6(1)(a) CPUT Regs
15 *R (on the application of Surrey Trading Standards) v Scottish and Southern Energy plc* [2012] EWCA Crim 539
16 Forgery Act 1981

3. Contracts
17 Condition 22.8 SLC
18 Condition 22.1 SLC
19 Condition 22.7 SLC
20 Condition 22.4 SLC
21 Condition 25A.2 SLC
22 Condition 27.2 SLC
23 Condition 14 SLC
24 Condition 22.4 (a) SLC
25 Condition 22D SLC
26 www.ofgem.gov.uk/simpler-clearer-fairer/what-and-when
27 Condition 22B.5 SLC
28 s62(2) CRA 2015
29 s68 CRA 2015
30 s69(10) CRA 2015
31 [2001] UKHL 52, [2002] 1 AC

32 *West and Another v Ian Finlay & Associates (a firm)* [2014] EWCA civ 316

Chapter 3

· ·

The right to a supply

This chapter covers:
1. Who is entitled to a supply (below)
2. Getting your electricity supply connected (see p29)
3. Getting your gas supply connected (p32)
4. Security deposits (p36)
5. Disruption of supply (p40)

This chapter assumes that you are legally responsible ('liable') for your fuel supply. Check Chapter 5 to ensure that you are in fact responsible for the supply.

---

1. **Who is entitled to a supply**

Electricity

Contract suppliers have a duty to offer a contract when they receive a 'request from a domestic customer'[1] and will supply electricity to you if the contract is accepted. Normally it is quite clear that you are requesting a supply but, to ensure it is treated as valid, your request should include:

- details of the premises to be supplied;
- the day on which the supply should commence;
- maximum power to be supplied, if this differs from that normally required by an ordinary domestic customer;
- the minimum period for supply;
- any reference to a continuing supply already established at the premises (where relevant).

Under Standard Licence Condition (SLC) 22, the supplier must offer to enter into a domestic supply contract with you as soon as is reasonably practical. The duty to provide a supply is only enforceable by Ofgem because it is contained in the supplier's licence, not in the Electricity Act (see Chapter 14). The contract must be in a standard form containing all the terms and conditions, including the price and your right to terminate the contract (see p30).

If you accept the offer of a contract, the supplier must provide, and continue to provide, a supply of electricity until the contract is properly terminated, subject to certain exceptions detailed below. If the supplier fails to fulfil its obligations in the contract, this is a breach of contract that may give rise to legal remedies, including a right to compensation.

It is worth noting that the SLCs for contract suppliers make no mention of 'occupier' or anything else connected with your right to occupy the place where you want a supply of electricity. The term **'occupier'** is not defined in the SLCs but covers any person who occupies any premises legally, whether paying rent or some other charge, or paying nothing (see p27 for squatters).[2]

A small minority of customers are covered by **tariff suppliers** that have different legal provisions. Tariff suppliers' duties to supply are set out in the Electricity Act 1989, whereas contract suppliers' duties with their individual tariffs are set out in the licence granted to them by Ofgem. Tariff suppliers have a statutory duty to supply you if:[3]

- you are an owner or occupier of the premises; *and*
- you request a supply by giving notice in writing.

Relatively few customers remain who are subject to former tariff arrangements. Former tariff customers fall into the category of having their electricity supplied in accordance with the relevant legislation, namely the Electricity Act 1989 and regulations made under it.

There is, however, provision for an alternative status for these old tariff customers. This is known as a 'special agreement' under section 22 of the Electricity Act 1989. The terms and conditions which bind both you and the tariff supplier are the terms of the agreement rather than those under the Act. Under a tariff supplier's licence, a special agreement is referred to as a 'contract', so that a special agreement must be a designated supply contract and must conform to the licence conditions covering the form and content of designated supply contracts (see p30).

There are two circumstances in which the question of a special agreement might arise.

- A supplier has the discretion to grant you a special agreement if you ask for one when you give notice requiring a supply.
- You could be required to enter into a special agreement by the supplier if it was 'reasonable in all the circumstances'.

In practice, most domestic supplies are now by way of contract and reference may be made to the SLCs.

Guaranteed standards of performance

Electricity distribution companies are subject to guaranteed standards under the Electricity (Connection Standards of Performance) Regulations 2015.

If the distributor fails to meet the standards, you are entitled to receive a payment. The size of the payment depends on the situation and the length of delay.[4] For example, if a quotation for a connection is not provided within five working days, the supplier is liable to pay you £15 for each day, including the day on which the quotation is provided.[5] (Various sums may become payable where a supplier fails to provide a schedule of works and starting times for different types of connection, within various time periods.[6])

These payments can be made direct to you or via your electricity company.

Disputes as to whether compensation is payable may be referred by Citizens Advice consumer service to Ofgem.[7] Ofgem must determine the dispute within 80 days unless satisfied that special reasons apply for extending the period.[8] Ofgem must issue a timetable and the list of documents received for determining the dispute and may also hold an oral hearing.[9]

Exceptions to the minimum standards of performance

In a number of situations the minimum standards of performance do not apply.[10] These include where:

- you inform the supplier or operator that you do not wish any action to be taken or you agree another course of action;
- you give information at the wrong address or outside the hours that the supplier or operator has specified;
- it is not reasonably practicable for the supplier or operator to act in a prescribed time owing to severe weather conditions, an industrial dispute or the action of a third party;
- the supplier or operator has been unable to gain access to premises;
- you have failed to pay the relevant charge after receiving a notice or where you have committed a criminal offence;
- there are exceptional circumstances beyond the control of the supplier or operator.

Liability for danger and harm arising from interruptions in supply

Suppliers and distributors are bound by the Electricity Safety, Quality and Continuity Regulations 2002[11] which lay down a duty to prevent danger from any works or equipment used in supplying electricity.

'Danger' includes danger to health or danger from electric shock, burn, injury or mechanical movement, or from fire or explosion, as a consequence of the generation, transmission, transformation, distribution or use of energy. It covers dangers to both humans and domestic and farm animals.[12] Generators, distributors and meter operators are placed under a wide duty to ensure that all electrical equipment is sufficient for the purposes and the circumstances in which it is used and that it is constructed, installed, used and maintained so 'as to prevent danger, interference with or interruption of supply' so far as is reasonably practicable.[13] A special duty is imposed on equipment such as a meter situated in your home.[14] A duty of co-operation between generators, distributors and suppliers is also imposed under the regulations.[15]

A distributor may also be liable for acts and omissions that amount to negligence, though technical and expert evidence may be needed to establish this – eg, in the case of a fire. Mere breaches of the regulations applying to distributors do not give rise to action for breach of statutory duty[16] but where harm results from negligence an action for damages may be available. The Divisional Court ruled:

> ...where there has been a breach of the Regulations by a given distributor, that does not mean that it was culpably negligent; however, such a breach may point to a breach of the duty of care although in practice evidence which goes beyond the mere breach may well be required to establish negligence. A simple failure consistently to perform or discharge a statutory duty with no reasonable explanation or justification therefore may provide grounds for a claim in negligence.[17]

However, it is still necessary to prove that the breach of the regulations actually caused the harm or damage. If the harm would have occurred even if the duty had been carried out there is no sustainable claim.

Gas

If you want a supply of gas, you have rights and obligations similar to those you have for electricity.

You have the right to be connected to the gas network by a gas transporter (see Chapter 1). Gas mains and service pipes are owned by gas transporters. There are a number of gas transporters, but National Grid Gas (previously Transco) is the main one. A gas transporter has a statutory duty to connect your premises to the gas mains if:

- you are the owner or occupier (an **'occupier'** is a person who occupies any premises legally, whether or not s/he pays rent or some other charge); *and*
- the premises are within 23 metres of the nearest gas main.

Where there is an existing domestic supply, you obtain a contract by contacting the gas supplier. Under SLC 22 a gas supplier must offer to enter into a gas supply contract with you after receiving a request. The offer must be made 'within a reasonable time' of receiving the request.

A domestic supply contract or a deemed contract must include:[18]
- the identity and address of the supplier;
- the services provided (including any maintenance services) and any service quality levels to be met;
- if a connection is required, when that connection will take place;

- the means by which up to date information on all applicable tariffs and any maintenance charges may be obtained;
- any conditions for renewal of the contract.

If there is no existing supply, you must inform the transporter in writing that you require a supply of gas at the premises concerned (there will normally be a standard form). You are charged for the costs of connecting your premises to the network (see p33) unless you choose to have independent contractors do the work for you (see p30).[19]

All gas suppliers are under an obligation to supply gas under a contract to new customers (ie, owners or occupiers who request a supply) whose premises are already connected to the gas mains either directly or by a service pipe. The 'obligation to supply' is a condition of each supplier's licence. You cannot enforce the obligation to supply without the help of Ofgem because it is contained in the supplier's licence rather than in legislation. All gas suppliers must publish the principal terms of the contracts available from them and bring them to the attention of customers.[20] These cover the terms and conditions under which gas is supplied and are regulated by conditions within the supplier's licence regulated by Ofgem, with the broader framework set by European Union (EU) Regulations and by consumer protection provisions (see Chapter 14). The situation is likely to change following the June 2016 referendum decision to leave the EU.

If you have been supplied under the terms of a contract initially, the supplier must continue to supply you until either the contract comes to an end or it is terminated. A gas supplier continues to supply gas under the terms of a deemed contract if your contract has come to an end (see p17).[21] A deemed contract must not:

- have a fixed-term period for the contract to run or have a termination fee;[22] *or*
- require you to give notice before you are able to change supplier, except where there is a change in ownership of the premises. The supplier must invite you to enter into a further contract to run immediately following the expiry or termination of your existing contract.

The obligation on a gas supplier to supply you with gas under the terms of a deemed contract could also arise if a supplier is ordered to do so by Ofgem. Ofgem has the power to suspend or revoke a supplier's licence. If this happens, or if a supplier is unable to continue to supply gas (eg, if it goes into liquidation), Ofgem can order an alternative supplier to supply you instead.

There are minimum standards for supply, distribution and reconnection of gas, under the Gas (Standards of Performance) Regulations 2005 and the Electricity and Gas (Standards of Performance) (Suppliers) Regulations 2015.

Obligation to complete a supply transfer within three weeks

Where you are seeking to transfer your supply of gas or electricity, the transfer must be completed within 21 days.[23]

Exceptions arise where:

- you request that the supply is completed at a later date; *or*
- you cancel the supply contract and notify the supplier that you do not wish the supply transfer to take place; *or*
- an existing supplier is entitled to block the transfer because of an unpaid debt; *or*
- the supplier does not have all of the information it requires to complete the transfer, having taken reasonable steps to obtain the information, and it cannot readily obtain that information from another source; *or*
- you are currently taking a supply of electricity through an exempt distribution system.

A supplier must ensure that all debts or charges of £500 or less can be assigned to a new supplier.[24]

Exceptions

There are situations where a supplier/transporter does not have a duty to supply or continue to supply. A supplier/transporter is entitled to refuse to connect a supply or to disconnect a supply which has already been given in certain circumstances – see p31 for electricity and p34 for gas.

Squatters

Squatting in residential buildings is a criminal offence under section 144 of the Legal Aid, Sentencing and Punishment of Offenders Act 2012. Suppliers are entitled to refuse to supply fuel to anyone committing an offence under section 144. Furthermore, an arrangement of a supply by a trespasser might be taken as evidence that an offence under the Act is being committed by demonstrating an intention to live in the building.

However, the offence is only committed where a person is deemed to be 'living' in the residential building and does not affect squatted premises which involve non-residential activities – eg, where a residential building is used other than for living purposes such as the storage of goods, commercial purposes, or cultural purposes such as concerts or exhibitions. Nor does it apply to occupancy of non-domestic dwellings such as former shops or pubs.

The legislation considers 'living' as meaning residing for any period of time, but much will depend upon how courts approach cases on their facts, as living at an address is not simply determined by the amount of time spent at a property.[25]

The precise scope of law has yet to be clarified on such matters as temporary occupation in cases where a person may have a settled home elsewhere or cases of mistake.

There may be other situations where a supply is possible without involving a person living in the building – eg, where a trespasser occupies a garden or yard or is living in a non-domestic part of a building such as a garage.

Persons who had a temporary previous lease or licence are specifically excluded from the offence of squatting[26] and remain entitled to a fuel supply.

Travellers

Local authorities are provided with guidance on practical aspects of site provision and management by the Department of Communities and Local Government and the Welsh and Scottish governments.[27] The guidance contains the standards for the supply of electricity, gas and water to sites. Standards vary according to the type of site provided.

The local council as landlord is responsible for the supply and is entitled to resell electricity, although its charges cannot exceed the maximum resale price for electricity (see p196).

Mobile home and caravan sites

Sites for mobile homes are licensed by local councils under the Caravan Sites and Control of Development Act 1960. Model standards for the provision of electricity and water are issued to local councils. Local authorities can decide what conditions, if any, to attach to caravan site licences. Guidance to local councils is in very broad terms and suggests that sites should be provided with an electricity supply sufficient to meet all reasonable demands of the caravans situated on them. There is no guidance for the supply of gas.

Always look at the provisions in the caravan site licence to determine if your site owner is obliged to provide a supply of electricity to your site. Site owners may resell electricity to you, but cannot charge more than the maximum resale price (see p196).

If there is no provision for electricity on your site, there is nothing to stop you from applying for a supply to be connected if you are the owner or occupier of a mobile home or caravan. However, note the possibility of significant connection charges.

Contact the National Association of Caravan Owners (www.nacoservices.com) for further advice.

2. **Getting your electricity supply connected**

Notice

To obtain a supply of electricity you must contact the supplier with your request. You can do this by writing a letter or by completing the supplier's standard application form. However, most suppliers do not always require written notice and will connect your supply if you telephone to request a supply or if you call into a customer service centre. You may wish to safeguard your rights by making your request in writing or by following up your telephone request in writing. Keep a copy of your letter in case a dispute arises.

Most application forms contain all the necessary details. If you are writing a letter but are not sure what maximum power is required, it should be sufficient to make clear that you want an ordinary domestic supply. If you have a preference, also specify what type of meter you would like and how you wish to pay for your supply (see Chapter 4).

If you choose not to give information about yourself (eg, about your previous address or creditworthiness), you may be asked for a security deposit (see p36). You do not have to give information about other people living in your home, but note that liability for the bill may be decided on who signs the application form or letter (see Chapter 5).

If your requirements on the application form are acceptable (ie, if the supplier is prepared to supply you on your choice of meter, method of payment or other terms and conditions), a contract supplier will offer you a contract that you can accept or reject.

When your supply is connected

If your home has been previously supplied with electricity, you should be given an appointment within two working days for your supply to be connected and a meter installed. If the appointment is not made within the specified time, you are entitled to automatic compensation of £65. If the appointment is made and not kept, you are entitled to automatic compensation of £65.

If your home has never previously had a supply of electricity, and you make a written request for an estimate of the charges of connection, this should be sent within five working days if the work is simple, or within 10 working days if the work is complicated. You are entitled to a compensation payment if the estimate is not sent within these time limits. For a contract supplier, connection times depend on how quickly you respond to its offer of a contract.

If you experience unreasonable delays in getting your supply connected, contact Citizens Advice consumer service.

Conditions of supply

Standard Licence Conditions (SLCs) apply to all suppliers of electricity. A supplier is under a duty to offer you a contract containing standard terms and conditions that comply with its licence conditions – eg, SLC 22.3 provides that fuel will not be supplied unless it is done so under a domestic supply contract or a deemed contract.[28] Once you sign a contract, you are legally bound by its terms, but if you think any are unfair or unreasonable, Ofgem has the power to stop a supplier enforcing any term which is incompatible with its licence conditions. See p15 for contractual terms and Chapter 14 for details about unfair terms.

Charges for connecting a supply

Connecting an electricity supply is carried out by a distribution network operator (DNO), which is licensed to distribute electricity through cables and provide connections to premises. Distributors are not responsible for meter reading or billing – your energy supplier does this.

You may be charged for the connection of a supply. Details of connection charges are available from the DNO's website.

To obtain a connection, you need to notify the DNO, within a reasonable time, of the details of the premises to be connected, the time the connection is required and (to the best of your knowledge) the maximum power to be supplied.

A DNO should provide you with a quotation for connection to its distribution system, but does not normally fit a meter until instructed to do so by your chosen electricity supplier. When providing a quotation, the DNO normally specifies that you need to nominate a supplier before connection takes place, and preferably before accepting the quotation. It is advisable to appoint and sign a contract with an electricity supplier at least 28 days before the date you want the electricity to flow.

Getting your supply connected

When accepting a connection from the DNO, you or your supplier are obliged to enter into a connection agreement. A '**connection agreement**' outlines the rights and obligations associated with the connection.

On connection, the DNO is obliged to maintain the connection for as long as required and to repair or replace any electrical lines or plants when necessary (except when you may be responsible for any damage to the equipment).

If you are dissatisfied with any aspects of connection, complain in the first instance to the company concerned (see Chapter 14).

Independent electrical engineers may be employed for certain electrical connection work. A list of independent electrical engineering companies can be obtained from a local DNO. Lloyd's Register operates the National Electricity Registration Scheme and a list of companies registered as competent for electrical connection work can be found at www.lr.org/en/utilities-building-assurance-schemes/uk-schemes/national-electricity-registration-scheme/.

Cables and wires running between your meter to your electrical appliances are not covered by any connection agreement with the DNO and the electricity supplier – a qualified electrician would have to install them for you. Customer protection equipment such as fuse boxes and switches are also not covered by the connection agreement.

When you can be refused a supply

Electricity suppliers may refuse to supply electricity, refuse to connect a supply to new premises, disconnect an existing supply, or refuse to reconnect a supply which has been disconnected. Disconnection for arrears is dealt with in Chapter 8.

You may be refused a supply for a number of reasons. Some of these reasons applying to tariff suppliers were set out in the Electricity Act 1989, but grounds for refusal are now set out in the SLCs for contract suppliers. All of the following reasons apply.

- You refuse to take a supply on the terms offered.
- You have not paid your bill for any electricity supplied, standing charges, meters and any connection charges within 28 working days of the date of the bill (see p89). You are entitled to two working days' notice of disconnection. Your supply may only be disconnected in relation to the premises where the debt arose.[29] Every supplier is required to have a code of practice on payment of bills, including procedures to deal with customers who have difficulty paying (see p89).[30] The terms of your supplier's code could protect you from disconnection by setting out alternatives.
- You did not pay your bill for any of the above charges at your previous address. The supplier may refuse to connect a supply at your new address. The supplier is not entitled to payment of your arrears from the next occupier of your previous address. Similarly, you cannot be held liable for debts left by previous occupiers of your new address.
- You have not paid a security deposit within seven days of being sent a notice requiring you to do so and the requirement of a security deposit is reasonable in the circumstances.
- You refuse to accept a supply under a 'special agreement' under the Electricity Act 1989.
- Your premises are already being supplied by another electricity supplier under arrangements which have not expired or been terminated.
- Supplying you with electricity would, or might be, unsafe – eg, because the wiring is in a dangerous condition.
- You refuse to take your supply through a meter.
- There has been damage or tampering to a meter and the matter has not been remedied (see Chapter 9).

- The supplier is prevented from supplying you by circumstances outside its control – eg, if it has been prevented from laying cables because of extreme weather conditions.
- It is not reasonable in all the circumstances. This is a 'catch-all' provision. Most disconnections or refusals to supply will be on one or more of the grounds above, but this provision might be used if those grounds no longer apply – eg, you are no longer in arrears but the supplier is insisting that you can only be supplied through a prepayment meter for future consumption. If you refuse to accept a prepayment meter, the supplier must demonstrate that disconnection is the only reasonable alternative. The supplier cannot use this catch-all provision unless it first gives you seven days' notice of the intention to disconnect.

Where you have a prepayment meter, in addition to providing you with information about its alternative cheapest tariff, the supplier must inform you that, as a domestic customer with outstanding charges, you may be able to change supplier by agreeing with a new supplier that the outstanding charges are assigned.[31]

Note: suppliers are not entitled to disconnect for alleged non-payment if the amount is genuinely in dispute.

---

3. Getting your gas supply connected

If you move into a home which is not physically connected to the gas mains network, you will need to arrange a supply. There are three ways to get connected to a gas supply.

- Using a gas transporter who is licensed to supply gas through pipes and is under a duty to provide a gas connection where it is economical to do so. You are charged for the connection costs (see p30). There is also a charge if you ask the transporter to lay any pipes which are needed. In addition, you may be asked to pay a security deposit to the transporter.
- Through a licensed gas supplier who can arrange for pipes to be laid by either the local gas transporter or an independent contractor. The gas supplier can pass on the charge for providing the connection and the pipework. This charge may include an arrangement fee.
- Through a qualified independent engineer installing pipes between a meter and a gas appliance. Once your home is physically connected to the supply network, the gas transporter becomes responsible for the maintenance of the pipe. Ownership of the pipe concerned is transferred to the transporter.

If your home is already connected to the gas mains network, you need to enter into a contract with a gas supplier for your supply of gas. You do not have to use the supplier which previously supplying the premises, although you may be deemed to have a contract with it if you do nothing about it (see Chapter 5). If you do not know who the current supplier to your premises is, you can find out through the M number (see p80) helpline (tel: 0870 6081 524). Normally, the shipper and the supplier are the same company. If not, you are referred to the shipper, who can tell you who the supplier is. See also Chapter 2 for details of contracts for the supply of gas.

When your supply is connected

In practice, the gas supply often remains connected after the previous occupier moves out, so there is often no interruption to the supply. If the supply of gas has been disconnected, the gas supplier which you have chosen arranges a date to reconnect. It must do this as soon as reasonably practicable. If the meter has been removed, the supplier arranges with the transporter for a meter to be installed or may itself provide you with a meter. If there is already a meter in place, the supplier will make arrangements with the owner of the meter (usually National Grid) for the existing meter to remain in place. This does not apply when the meter in place is not suitable.

The relevant key standards state that when you request a gas supply:
- if a survey visit is required, contact is made within two working days to arrange an appointment which will be within three working days, or later if you requested;
- following such a visit, a quotation for providing a supply is sent within five working days of the visit if the property is adjacent to a public highway in which there is a suitable gas main, or otherwise within 20 working days;
- if no visit is required, a quotation is sent within five working days of receipt of the enquiry.

In the event of an unreasonable delay, take the matter up with Ofgem. Lloyd's Register operates the Gas Industry Registration Scheme on behalf of the UK Gas Transporters. For further information see www.lr.org/en/utilities-building-assurance-schemes/uk-schemes/gas-industry-registration-scheme/.

Charges for connecting a supply

You should not normally be charged for the connection of a gas supply when you are taking over the supply at premises that are already connected to the gas network. An exception is where the connection has been capped for over 12 months and a new connection is needed.

A gas transporter charges for connecting your premises to the gas network for the first time. You may be charged for all work done on your home and land and for any pipe which has to be laid, although the first 10 metres of the pipe that is not on your property is covered by the gas transporter. For domestic premises within 23 metres of a relevant gas main, a transporter is obliged to connect premises and provide and install the necessary assets for connecting the premises.[32] For premises further than 23 metres from a main or consuming more than 2,196,000 kWh, the gas transporter quotes a price for connection. All work to connect this type of premises is chargeable.

Potential customers may face high connection charges, particularly if a new supply is required some distance away from the gas mains network or if costs cannot be shared between a number of new customers. Information and quotations can be obtained from National Grid. Any charges should be checked closely to see if the expenditure is reasonably incurred; it may be possible to contest some charges.

Charges are based on National Grid recovering the cost of laying new mains within a five-year period, less a discount reflecting the anticipated revenue from the new customers. You may be charged if your gas main is less than five years old at the time that you ask for a gas supply. You may be asked to finish paying for the costs of having the supply put in, but the extra charge only applies if:

- the amount of the charge is no more than anyone previously supplied from the main has been charged; *and*
- the transporter has not yet recovered the full cost of the main; *and*
- the transporter has supplied you with any information you reasonably requested concerning the cost of the main, the date it was laid and how much has been paid by previous consumers.

This charge does not apply if you are an owner or occupier who has paid contractors to connect the supply.

Ofgem has a duty to resolve disputes about connection charges.

When you can be refused a supply

A gas supplier may refuse to connect a gas supply to a new address, may cut off an existing supply or refuse to reconnect a supply which has been disconnected. A gas transporter may refuse to connect your premises to the gas supply network and may also disconnect your supply in a number of circumstances. Disconnection for arrears is dealt with in Chapter 8.

A **gas supplier** may cut off your supply in the following circumstances.

- You do not pay your bill within the 28 days following the date of the bill (see p89). You are entitled to a minimum of seven days' notice of the supplier's intention to disconnect. **Note:** there is no right to disconnect when the bill is genuinely in dispute. You may be protected from disconnection by conditions contained in your supplier's licence.[33]
- You change your supplier and you owe money to your previous supplier, in which case the previous supplier can assign some of your debt to your new supplier.[34] The new supplier cannot refuse debt assignment and will collect the debt through a prepayment meter. However, the new supplier is unlikely to

cut off the supply. What is termed **'debt blocking'** – the refusal to take on a debtor with a prepayment meter as a customer by a new supplier – should not take place if a debt is £500 or less, following the debt assignment protocol (see p105).[35] You are entitled to a minimum of seven days' notice of the new supplier's intention to disconnect. The new supplier may also refuse to connect in these circumstances if your previous supplier cut off your supply and is still entitled to keep your supply cut off. The relevant debt does not include any sums for which you had already been billed by the previous supplier by the date you transferred to the new supplier and which you had failed to pay within 28 days. It includes any subsequent amounts for which you are billed by the previous supplier, providing you have failed to pay within 40 days. The previous supplier is only entitled to assign the debt if it has given you 14 days' notice of its intention to do so. If you are a credit meter consumer, you can ask for debt assignment, which is at the discretion of suppliers. If the debt is assigned, disconnection may arise from non-payment. **Note:** there is no right to disconnect when the bill is genuinely in dispute. You may be protected from disconnection by conditions in your supplier's licence (see p117).

- You do not pay a reasonable security deposit or agree to accept a prepayment meter within the seven days following the supplier's request for a deposit. A supplier's right to request a security deposit is a condition of the supplier's licence.
- You fail to take your supply through a meter.
- You fail to keep a meter belonging to you or to someone other than the gas supplier or transporter in proper order.
- You intentionally damage or interfere with gas fittings, service pipes or meters (see Chapter 9).
- You do not/no longer require a supply of gas.
- You do not/no longer require the use of meters or other gas fittings belonging to the supplier/transporter. You are entitled to 24 hours' notice.
- Your supply has been reconnected without the consent of the supplier.
- Supplying you with gas would, or might, involve danger to the public.
- A gas transporter or another gas supplier has disconnected your supply and is under no obligation to reconnect your supply.
- A gas shipper has prevented the transfer of gas to your premises.
- A supplier's ability to supply its customers would be significantly prejudiced if it were to offer you a supply.
- There are circumstances beyond the supplier's control.[36]

A **gas transporter** may refuse to connect your premises to, or may disconnect your premises from, the gas supply network in the following circumstances.
- Your premises are not within the transporter's authorised area.
- Your premises are not close enough to a gas main – ie, the premises are not within 23 metres of the transporter's gas main or could not be connected by a service pipe to a transporter's gas main.

- A transporter asks you to install a meter as near as possible to its main and you refuse. This applies when:
  - gas was not previously supplied to your premises by the transporter; *or*
  - a new/substituted pipe is required; *or*
  - the meter is to be moved.

  Note that the transporter may permit you to install a meter in alternative accommodation or in an external meter house, but this discretion lies with the transporter.
- You use gas improperly or deal with gas so as to interfere with the efficient conveyance of gas.
- The transporter is concerned to prevent the escape of gas or it suspects there may be an escape of gas.
- You fail to take your supply through a meter.
- You do not pay a reasonable security deposit. A gas transporter can request reasonable security for the initial connection of the supply.[37]
- You fail to keep a meter belonging to you or to someone other than the gas supplier or transporter in proper order.
- You intentionally damage or interfere with gas fittings, service pipes or meters (see Chapter 9).
- You do not/no longer require a supply of gas. You are entitled to 24 hours' notice.
- You do not/no longer require the use of meters or other gas fittings belonging to the transporter. You are entitled to 24 hours' notice.
- The transporter is prevented from connecting you or maintaining your connection by circumstances not within its control.
- Supplying you with gas would, or might, endanger the public.
- A pipe laid by the owner or occupier of the premises is not fit for the purpose.

4. **Security deposits**

What is a security deposit

A '**security deposit**' is a sum of money requested by electricity or gas suppliers as a condition of providing a supply. Gas transporters can also ask for a security deposit as a condition of connecting your premises to the mains network.

Deposits are held separately from customers' normal accounts and are used to offset costs for the supply of electricity or gas, usually following a disconnection. Deposits should not be regarded as a credit payment towards future bills. Note, however, that the licence conditions for gas suppliers appear to allow the possibility that a supplier may use some or all of your deposit to reduce your gas bill. See p39 for the return of a deposit.

When you can be asked to pay a security deposit

Electricity

Rules for security deposits are set out in Standard Licence Condition (SLC) 27.
Electricity suppliers may ask for a security deposit if:

- you refuse to take a supply through a prepayment meter; *or*
- it is not practicable to install a prepayment meter; *and*
- it is reasonable in all the circumstances to do so (a tariff supplier's power is expressed as the right to ask for reasonable security, which amounts to the same thing).

When a contract supplier asks for a security deposit, it must inform you of when it will be returned (see p39) and of the power of Ofgem to determine any dispute about the deposit. As an alternative, action through the county court (or sheriff court in Scotland) can be used to recover a deposit which is owed to you and which a supplier refuses to refund (see Chapter 14).

If you are a new customer, you may be routinely asked for a security deposit by electricity suppliers, particularly if:

- you refuse to provide information about previous addresses, or you cannot demonstrate a satisfactory payment history at a previous address and you do not otherwise provide sufficient information about your creditworthiness; *or*
- you have been assessed as having a poor credit rating; *or*
- you are in short-term accommodation. You should not be treated as being in short-term accommodation if you are a secure or assured tenant.

Suppliers cannot insist on both a prepayment meter and a security deposit (see p40).

Your supplier's code of practice on the payment of bills should include a statement of its policies on security against the non-payment of future bills. The code should also state if and how policy differs for new and existing customers and what, if any, credit-vetting procedures are used. It should indicate the steps you need to take to improve your creditworthiness or to ensure that security is no longer needed.

Security deposits may also be required from existing customers if:

- your payment plan has broken down. You will almost always be able to have a prepayment meter as an alternative (see p40). Remember that the requirement for a security deposit must be reasonable. If the reason your payment plan broke down was that you could not afford it, you may be able to negotiate another payment plan instead of either having to pay a security deposit or having a prepayment meter installed (see Chapter 7); *or*
- theft, tampering or damage to meters/equipment has occurred (see Chapter 9).

It is unlawful for a supplier to discriminate unduly in how it supplies customers, including in respect of security deposits – eg, if people who live on a particular estate are asked for a security deposit. Contracts are individual agreements and a supplier should avoid discriminating against any particular class of customer. A complaint about this may be made to the supplier and also to the Ombudsman. Details of the codes may be referred to in court if a case leads to legal action.[38]

Gas

Gas suppliers can, under the terms of their licence, incorporate demands for security deposits in their contracts and 'deemed contracts' (see p17). Deposits may be cash deposits, or a secure method of payment such as a direct debit or a prepayment meter. Conditions in the supplier's licence will limit the circumstances and amounts of deposits that may be requested. Suppliers cannot ask for a security deposit if you agree to a prepayment meter, unless your conduct makes it reasonable to ask for a deposit.

A gas supplier must not require payment of a deposit where 'it is unreasonable in all the circumstances of the case to require that customer to pay a security deposit.'[39] The wording indicates that the supplier is required to consider the individual circumstances of a customer and cannot apply a blanket policy of imposing deposits upon particular classes of customers or certain areas. The deposit must not exceed an unreasonable amount.[40]

Your supplier's code of practice on the payment of bills will provide a statement of its policies. All suppliers are subject to the same obligations in respect of security deposits. If you do not provide the security requested, the supplier may refuse to connect your supply, if you are a new customer, or disconnect your supply, if you are an existing customer. Security can mean:
- you pay a cash deposit; *or*
- you join a gas payment plan; *or*
- a prepayment meter is fitted.

Suppliers may ask for security if:
- you live in short-term accommodation. You should not be treated as being in short-term accommodation if you are a secure or assured tenant; *or*
- you have a poor payment record at your present or last address; *or*
- you are a new customer and you do not give proof of your identity or your last address.

A gas transporter is also entitled to ask for a security deposit if connecting premises to the gas pipe network where the premises are no more than 23 metres from the gas main and it will be laying the pipes needed for the connection. The transporter may refuse to supply and lay the pipe if you fail to pay the security requested.

Amount of deposit

Electricity

For a contract supplier, the deposit should be 1.5 times the value of the average quarterly consumption of electricity reasonably expected at the premises. The amount can only be more than this if that is reasonable in all the circumstances.[41]

The amount of deposit required may be increased if the existing security has become invalid or insufficient. The amounts actually requested will vary between suppliers.

Gas

Gas suppliers' licence conditions state that the amount of a deposit must not exceed a reasonable amount.[42] If you consider the amount unreasonable, the amount should be referred to Ofgem. What is reasonable requires a consideration of all relevant facts, including your income and capital and personal factors such as disability and previous payment record. Previously, it was considered that gas suppliers should only request a deposit of a maximum of 1.5 times the value expected in quarterly consumption.

Disconnection if you do not pay a security deposit

An electricity supplier, gas supplier or gas transporter may refuse to connect, or may disconnect, your supply if you fail to pay the requested deposit within seven days of being billed. Your supply may remain disconnected for as long as you refuse to pay the amount requested.

If you cannot afford to pay a reasonable security deposit for a gas supply straight away, your supply will only be disconnected as a last resort if a prepayment meter cannot be installed (or is refused) or Fuel Direct is not available to you. See Chapter 8 for more on disconnections.

A gas supplier is not entitled to withhold the supply or threaten to disconnect for any amount of security deposit which is genuinely in dispute. This applies if you dispute the amount of, or the need for, a security deposit and breaches of the standards of performance.

Ofgem and the Energy Ombudsman may consider and resolve disputes over security deposits. See Chapter 14 for how to resolve a dispute.

Return of security deposits

Electricity

Contract suppliers must repay any deposit:
- within 14 days where, in the previous 12 months, you have paid all charges for electricity within 28 days of each bill being sent to you; *or*
- as soon as reasonably practicable, and in any event within one month, where you have stopped taking a supply from that particular supplier and have paid all outstanding charges.

Where there is a failure to return a deposit, a small claim could be commenced through the civil courts (see p230).

Gas

Security deposits held by a gas supplier under the terms of a contract or deemed contract must be returned to you if, for a continuous period of 12 months, you:

- pay your bills within 28 days of their being issued; *or*
- otherwise comply with the terms in respect of payment under the terms of your contract.

The deposit must be returned within two months of this period, unless it is reasonable for the deposit to be retained due to your conduct.

If you have stopped taking a supply within a period of 12 months and paid all dues, check the terms of your contract in relation to the deposit.

Interest on security deposits

Electricity

Interest is payable at simple interest rates on every sum held for more than one month by an electricity supplier. The rate of interest is the base rate of Barclays.[43]

Gas

The provision that gas suppliers should pay simple interest on any deposit held for more than one month is not included in SLC 27. To pursue a claim for interest would require court action on a claim for the return of the deposit.[44]

Disconnection and reconnection costs

Electricity suppliers may demand the reasonable expenses of both disconnection and reconnection as well as payment of a security deposit prior to reconnecting the supply.

A gas supplier may demand the expenses of disconnection and reconnection if it has disconnected for failure to pay a deposit. The supplier is not entitled to payment where the amount of a deposit is genuinely in dispute.

Alternatives to security deposits

- Suppliers routinely accept **payment plans** (see p95) as acceptable alternatives to cash security deposits.
- Some electricity suppliers accept **guarantors** as an alternative to a security deposit. Potential guarantors should be aware that if the bill is not paid, they are liable for the debt and could be pursued for the debt through the courts.
- Suppliers are not entitled to a security deposit if you are prepared to have a **prepayment meter** and it is reasonably practical for the supplier to provide one. Suppliers can take into consideration the risk of loss or damage to a meter in deciding whether a prepayment meter can be offered. You may be asked to pay a security deposit and have a prepayment meter if this is reasonable as a result of your conduct – eg, if there is evidence that you may damage the meter.

5. **Disruption of supply**

Suppliers may be liable to pay compensation for any disruption in the supply of electricity or gas. Such disruption can arise in many ways such as severe weather, equipment failure and vandalism. Generators and distributors are also under a duty to avoid interferences and interruptions of supply caused by insufficient clearance between overhead lines and trees or other vegetation.[45] Various options are open to you if you suffer a loss of power, including a claim for compensation under the Electricity (Standards of Performance) Regulations 2015 or a civil action for damages through the courts. In many cases, the claim lies against the electricity distribution network operator (DNO).[46]

DNOs have to meet guaranteed standards of performance for restoring supplies to customers. The basic principle is that a payment is paid where a supply to your home is interrupted as a result of a failure or fault in or damage to a distributor's system and not restored within a set time period. Further payments are payable for each subsequent 12-hour period in which power is not restored.

If your electricity supply fails during normal weather conditions because of a problem with the distribution system, the DNO should restore it within 18 hours of becoming aware of the problem. If the DNO fails to do so and you make a valid claim within three months, you are entitled to £75 and a further £35 for each additional 12 hours you are without supply.[47]

A longer period applies for larger scale interruptions of power, affecting large numbers of customers. If the incident involves 5,000 customers or more, the DNO is required to restore supply within 24 hours. If the DNO fails to do so and you make a valid claim, you are entitled to £75 and a further payment of £35 for each additional 12-hour period that you are off supply up to a maximum of £300.[48]

If your electricity supply fails because of a problem on the distribution system due to severe weather, generally, if a supply is not restored within 24 hours, you are entitled to £70 and a further £70 for each additional 12 hours you are without supply to a maximum payment of £700.[49]

The scheme does not apply to an island where the supply is provided via a line situated on or under the seabed.[50]

Ofgem will look into cases of power cuts.

A claim can be started by writing directly to the supplier with details of the disruption to supply. The letter is the key document to begin the process and should set out details of the dates and times when the loss of power occurred (so far as it can be identified) and any particular consequences it has had on members of the household concerned. A reasonable time limit of 14 days should be given to the supplier to respond. Contact Ofgem about a claim for sums payable under the regulations arising from disruption of supply if the company concerned has already offered a payment, particularly if the payment (known as an ex gratia payment) is lower than the sum suggested by the regulations. Equally, amounts may be higher where there has been special damage which has arisen from the loss of power.

Energy Ombudsman

The Energy Ombudsman (see Chapter 14) can act in claims arising out of power cuts. It can ask the company to take practical action to resolve a dispute and, in some cases, make a financial award.

Court action

As an alternative to seeking compensation from a supplier, if you have lost either electricity supply or gas under contract, you may bring a claim under contract or negligence through the civil courts (known as 'delict' in Scotland).

Claims may be brought in relation to the terms of the contract or with reference to section 14 of the Supply of Goods and Services Act 1982, which puts an implied term into every consumer contract for services that the service will be provided with a reasonable degree of competence and skill. Where there is a failure in the service, the provider may be liable to pay compensation for breach of this implied term.

Sums awarded by a court for nuisance and inconvenience arising from disruption of supply are likely to reflect those set out in the regulations. In addition, any losses which flow directly from the breach of supply and which are reasonably foreseeable as a result of power loss may also be recoverable. For example, these might include the cost of spoiled food in a freezer where power supply has been lost or where damage has been caused to a computer hard drive by a loss of power.

If a failure to supply power results in serious loss or damage or personal injury, legal advice should be taken on a claim.

It is possible that in extreme weather cases, energy companies will claim the benefit of a common law defence known as 'Act of God'. If such a claim is raised, seek specialist advice.

Liability for damage arising from faulty connection

If damage arises from a faulty connection (or reconnection), the energy company and/or its sub-contractor are liable. An energy company is liable for the negligence of its sub-contractor under a principle known as 'vicarious liability'. Both the supplier and its sub-contractor could be treated as defendants in any claim, pleading the negligence of the sub-contractor and negligence on the part of the supplier in selecting an unsuitable sub-contractor. Seek legal advice if the claim is substantial or personal injury has been caused.

Notes

1. Who is entitled to a supply
1 Condition 22 SLC
2 *Woodcock v South Western Electricity Board* [1975] 2 All ER 545
3 ss16 and 64 EA 1989
4 Sch 1 E(CSP) Regs
5 Reg 5(2) and Sch 1 E(CSP) Regs
6 Sch 1 E(CSP) Regs
7 Sch 2 E(CSP) Regs
8 Sch 2 para 2(1) and (2) E(CSP) Regs
9 Sch 2 para 2(2) and (5) E(CSP) Regs
10 Sch 2 E(CSP) Regs
11 SI 2002 No.2665, as amended by the ESQC Regs
12 Reg 1(5) ESQC Regs
13 Reg 3(1) ESQC Regs
14 Reg 24 ESQC Regs
15 Reg 4 ESQC Regs
16 *Morrison Sports Ltd and others v Scottish Power* [2010] UKSC 37, [2010] 1 WLR 1934, [2011] UKSC 1
17 *Smith and others v South Eastern Power Networks plc and others* [2012] EWHC 2541, per Akenhead, J
18 Electricity and Gas (Internal Markets) Regulations 2011 No.2704 as amended by the Electricity and Gas (Internal Markets) Regulations 2014 No.3332 from 14 January 2015
19 s10(1)(b) GA 1986; see also condition 4B Gas Transporters Standard Licence Conditions
20 Condition 23.1 SLC
21 Conditions 7 and 23.2 SLC
22 Condition 7.6A SLC
23 Conditions 14.4 and 14A SLC; Sch 7 Part 3 and Sch 8 Part 4 Electricity and Gas (Internal Markets) Regulations 2011 No.2704 as amended
24 Condition 14.6 SLC
25 *Doncaster Borough Council v Stark and Another* [1997] CO/2763/96 5 November 1997, Potts, J; *Frost (Inspector of Taxes) v Feltham* [1981] 1 WLR
26 s144(2) Legal Aid, Sentencing and Punishment of Offenders Act 2012

27 Department of Communities and Local Government, *Planning policy for traveller sites*, September 2015; Welsh Government, *Designing Gypsy and Traveller Sites*, May 2015; Scottish Government, *Improving Gypsy/Traveller Sites – Guidance on minimum site standards and site tenants' core rights and responsibilities*, May 2015

2. Getting your electricity supply connected
28 Condition 22.3 SLC
29 Sch 4 para 2(1) UA 2000
30 Condition 27.5 SLC
31 Condition 31E.4 SLC

3. Getting your gas supply connected
32 Condition 4B Gas Transporters Standard Licence Conditions
33 Condition 27 SLC
34 Condition 14 SLC
35 Condition 14.6 SLC; Ofgem debt assignment protocol for prepayment meter customers letter, 12 May 2015
36 Condition 22.5 SLC
37 s11 GA 1986
38 *Laverty and others v British Gas Trading* [2014] EWHC Ch 2721

4. Security deposits
39 Condition 27.3(b) SLC
40 Condition 27.4 SLC
41 Condition 27 SLC
42 Condition 27.4 SLC
43 Condition 27 SLC
44 s69 County Court Act 1984

5. Disruption of supply
45 Reg 20A ESQC Regs
46 Regs 4-10 E(SP) Regs
47 Reg 5(2)(a) and (b) and Sch 2 E(SP) Regs
48 Reg 6(2) E(SP) Regs
49 Reg 7 E(SP) Regs
50 Reg 9(3) E(SP) Regs

Chapter 4

· ·

Meters and methods of payment

This chapter covers:
1. Standing charges and tariffs (below)
2. Types of meters (p47)
3. Payment methods (p53)
4. Fuel Direct (p55)
5. Choosing how to pay (p55)

For more information on reading your meter, see Appendix 2.

---

1. Standing charges and tariffs

'**Standing charges**' are fixed charges which must be paid regardless of how much fuel you use. Suppliers make these charges to cover costs such as billing, meter reading, customer services, servicing meters and so on.

A '**tariff**' is the package of charges and conditions that a supplier offers you for providing electricity, gas or both.

Suppliers currently can offer four core tariffs for electricity and four core tariffs for gas – though in reality this means four each for gas and electricity for each meter type and payment method. **Note:** this is likely to change in the future, due to the Competition and Market Authority's review of the retail energy market, although changes are not likely to happen until 2017.

All tariffs are structured in the same way and comprise a standing charge and a unit rate (or rates, for time of use tariffs). Note that although the structure of all tariffs is the same, suppliers can choose to offer tariffs where the standing charge is set at zero.

Every tariff has a specific name. It's important to know the name of your tariff, particularly if you want to shop around for a better deal. The name of your tariff is shown on your bill and on your annual statement. Your supplier must tell you, via your fuel bill and other communications, whether it has a cheaper tariff available and how much you could save by switching to it.

All suppliers should provide a **'tariff information label'** (TIL) which details the key terms and conditions for each tariff. Your bill/annual statement must also include a **'tariff comparison rate'** (TCR), to provide 'at a glance' information to help you compare tariffs. **Note:** if your gas or electricity consumption is much higher or much lower than average, the TCR is of limited use because it is based on average consumption levels.

Suppliers are not permitted to increase prices on, or make other changes to, fixed-term tariffs without your consent, though structured price increases that have been set out in advance and that comply with consumer protection law are allowed. The following exemptions also apply:
- an increase in price due to an increase in VAT;
- your payment method is changed because of your debt and/or failure to comply with contractual terms.

If you are on a fixed-term contract that is coming to an end, your supplier cannot roll you forward onto another fixed-term contract without your consent. Your supplier will give you a six-week period before your contract is due to end to make a decision about your preferred tariff and supplier.

Prepayment meters

Historically, people using prepayment meters paid more for their energy than customers using alternative payment methods. However, there have been active steps to reduce discrimination in the cost of fuel, particularly between those who have prepayment meters and those who pay by direct debit or e-billing/paperless billing. This is included in Standard Licence Condition 25A, which provides that suppliers must ensure that they do not discriminate between groups of domestic customers.[1] Fuel suppliers must offer a range of payment methods but must justify price differences applied on the basis of payment method. Most prepayment meter customers now pay the same, or slightly less, for their fuel than customers on quarterly billing. However, some prepayment meter customers were previously on a tariff with no standing charge. Although this often meant that they were paying for fuel at a two-tier rate, many felt that they benefited during warmer weather because they did not use their heating and did not have to top up their meter as often. Now that a standing charge applies to all tariffs, some prepayment meter customers have found that they have accrued debt because of a build-up of standing charges.

Warm Home Discount

The Warm Home Discount (WHD) was introduced initially as a four-year scheme from 2011 to 2015 and has now been extended to 2021. It was set up to help low-income and vulnerable households meet their energy costs. This is a mandatory scheme, funded by the major fuel suppliers, which effectively replaced the social/

special tariffs previously available to a range of vulnerable and/or fuel poor customers.

It was originally intended to be a rebate on your electricity bill but, from 23 July 2016, energy suppliers can now provide the option to pay the rebate on the gas bill rather than the electricity bill, if you request this of your supplier.[2]

There are two main groups within the scheme.

- **Core group.** A yearly discount, of £140 in 2016/17, is made to the winter gas or electricity bill of each eligible customer. Eligibility alters each year, but is targeted at customers over pension credit (PC) age. In 2016/17, you are considered eligible if, on the qualifying date (10 July 2016 – the date is published annually on the gov.uk website), you were receiving the guarantee credit element of PC (even if you get the savings credit as well). The Department for Work and Pensions (DWP) data-matches its records with participating energy suppliers so, if you are eligible, you should receive the discount automatically.

- **Broader group.** Participating energy suppliers have discretion over the eligibility criteria for the broader group, but they are still required to target those in, or at risk of, fuel poverty. The eligibility criteria are subject to approval from Ofgem. The annual electricity discount for the broader group is £140 in 2016/17. There is no data-matching with this group so, if you think you are eligible, contact your energy supplier. The discount is awarded on a first-come-first-served basis.

Note: the £140 includes VAT so the rebate will show on your bill as a credit of £133 – ie, before VAT is applied.

If you have a prepayment meter and qualify for a WHD, contact your supplier to ask how you will receive the discount – eg, it may be paid as a bar-coded top-up voucher or a post office voucher.

Examples

May gets the guarantee credit of PC. The DWP writes to tell her that she will automatically get a £140 credit to her account.

Dev and Polly have a pay-as-you-go smart meter and apply for a WHD under their supplier's scheme because Dev gets income-related employment and support allowance and they have a four-year-old son. Their application is successful and £140 is credited to their account.

Rhys has a prepayment meter and applies for a WHD because his annual income is less than £16,190 and his fuel spend is more than 10 per cent of his income. His supplier sends him a £140 voucher to top up his account.

Miranda gets income-based jobseeker's allowance and personal independence allowance. She cannot get a WHD as her supplier does not participate in the scheme.

Note: WHD is due to be devolved to the Scottish government following the Scotland Act 2016. Details are not yet known.

For more information, see p170.

2. Types of meters

Meters are owned by meter operator companies contracted by suppliers to provide metering services to their customers. In many cases, meters are owned, checked and read by National Grid and the privatised electricity suppliers even if the fuel is supplied by another company, but more meter operator companies are being established. Metering remains the responsibility of the supplier, which is liable for the acts and omissions of its sub-contracting meter operator company.

The main types of gas and electricity meters supplied are: standard credit, variable rate credit, prepayment and smart.

Standard credit meter

Most customers have credit meters, with fuel supplied in advance of payment. Credit meters record consumption and are read periodically – usually twice yearly – by your supplier. There is no requirement under the standards of performance to read all meters regularly, but meters must be read at least once every two years under the supplier's licence. Estimated bills are sent for the rest of the year, with a customer reading correction facility available by phone or online.

Estimated bills are a frequent source of complaint. A succession of estimates can result in inaccurate billing, with you paying too much or not enough. This often leads to problems with arrears and the threat of disconnection. If you have problems with arrears as the result of a succession of estimated bills, see p93.

A bill is sent at the end of each billing period, after the meter has either been read, or was due to be read, or estimated. The price per unit of fuel does not vary according to the time of day or night the fuel is used when you use a standard credit meter. Appendix 2 describes how to read your own credit meter.

If you are of pensionable age or disabled, you could use the special meter reading facility under the Priority Service Register. Quarterly readings can be arranged for you and a special password service is available as protection against bogus callers. If you are registered disabled and it is difficult for you to reach or read your meter, your supplier may reposition it to a more convenient location free of charge.[3] Free safety checks on all gas appliances are also offered to a limited number of eligible households. Contact your supplier for information about the Priority Service Register services it offers.

Variable rate credit meter

Variable, or off-peak, electricity credit meters record different rates or 'tariffs' at different times of day or night. Night-time off-peak electricity usage is generally cheaper than on-peak day usage. The most common type of variable rate credit meter is known as Economy 7 in England and Wales and sometimes referred to as white meter in Scotland (see below).

Suppliers offer different systems, depending on your supply area, and these may change from time to time. Ask your supplier for information on the type of system it operates.

Economy 7/white meter

Economy 7/white meter is a scheme allowing you to pay for your electricity at two different rates or 'tariffs'. You need a special meter, usually an Economy 7 credit meter, but in some areas Economy 7 prepayment meters are available as well. A white meter is a similar type of meter which preceded Economy 7 meters. They are still very common in Scotland, but are being phased out elsewhere.

Electricity is charged for at two different rates per unit, with a lower rate at night. The daytime rate is charged at a higher rate than the standard rate for credit meters. The standing charge is often higher than for credit meters. The amount of the charges varies from supplier to supplier.

Consider changing to an Economy 7/white meter if you use electricity to heat your home and to heat water overnight. You may also be able to make savings in your fuel costs if you run electrical appliances (such as washing machines and tumble dryers) overnight, usually by using a timer to ensure the appliances operate within the optimum time band.

The higher standing charge and higher daytime rate may counterbalance any savings made if your night-time use of electricity is not large enough. Look carefully at the amount of electricity you use during the day and night, and at the rates, to establish if an Economy 7/white meter would save you money. Suppliers should have specialist staff to advise you.

Time of use (off peak) tariffs and meter clocks

There have been problems with time clocks for some meters which have left customers out of pocket. Consumer organisation Which? carried out an investigation in 2014 into faulty meter clocks.[4] Time-of-use tariffs, such as Economy 7 or Economy 10, offer a lower rate for electricity during specific hours (peak and off-peak times vary between tariffs, regions, meter types and seasons), but some meter clocks have been found to be wrongly set. The problem is largely down to the clocks on some meters not changing for GMT or BST at the appropriate time of year (some will change automatically) and, as a result, time-of-use tariffs are not charged correctly. Power cuts may also affect the clocks that control switchover times.

It is the supplier's responsibility to ensure metering equipment is correct so if you suspect there might be a fault, contact your energy company. Suppliers are not required to specifically check meter clocks, but Ofgem rules mean they must take reasonable steps to ensure accuracy in terms of the amount of electricity supplied.[5]

If necessary, the accuracy of the meter can be checked by the Regulatory Delivery office (part of the Department for Business, Energy and Industrial Strategy).

Prepayment meters

A substantial number of people have a prepayment meter – in the UK there are in excess of 4.5 million electricity meters and 3.4 million gas meters. Over 60 per cent of these meters were installed to recover a fuel debt.[6] Frequently marketed as a 'pay-as-you-go' budgeting method, there are several types of prepayment meters, including card and key meters. Some electricity prepayment meters can operate with Economy 7 and other variable rate tariffs.

Collecting arrears

Prepayment meters can be set to collect a fuel debt. They allow you to pay for your supply of fuel, a daily standing charge and extra for any arrears you owe. It is important to note that the settings for these charges operate on a regular, usually weekly, basis. If you are due to spend time away from your home (eg, on holiday or in hospital), ensure that your meter is 'topped-up' with enough credit to cover these charges.

If you are paying off a fuel debt via a prepayment meter, your supplier is meant to take into account your ability to pay when determining the weekly arrears recovery amount. If you feel the arrears repayment level is unaffordable, you may be able to negotiate with your supplier (especially if you can be considered vulnerable) to have the arrears recovery level set on a par with Fuel Direct (see p55) – ie, £3.70 a week.

Emergency credit

If your fuel runs out, you can use an emergency button on the meter to obtain a small amount of credit (typically worth £5). The next time you top up, the credit is used to pay for the emergency fuel – no more fuel is available until this has been paid.

Self-disconnection

With insufficient funds in a prepayment meter, you effectively disconnect yourself, rather than the energy supplier having to take steps to enforce any debt. Smart prepayment meters are being developed.

Information displayed on prepayment meters

Prepayment meters have a liquid crystal display which displays a range of information about how the meter has been set – eg, how much credit remains, the total outstanding debt remaining or the amount of emergency credit you have used.

Advantages of prepayment meters

- They can be useful as a budgeting aid, as they restrict your use of fuel according to your means. You are forced to become aware of your fuel consumption. This can be useful if your budget is limited, but you should also consider the risk of self-disconnection. Many customers choose to retain their prepayment meter as a budgeting aid even after arrears have been paid off.
- Ease of adjustment – the smart card or key reads your meter and conveys the information from your supplier. If your supplier agrees to change the setting, there is no need for a visit, as the card/key adjusts the setting of the meter the next time you charge it up and use it.
- Meters can be reset to pay off arrears as an alternative to disconnection.

Disadvantages of prepayment meters

- These meters should never be installed if you are at risk of leaving appliances turned on after the money has run out, or are incapable of operating the meter to obtain credit or emergency credit. They also should not be installed if you cannot obtain the tokens, cards or keys to operate them.
- There are hidden costs. If you cannot afford to buy much fuel at any one time, you will need to make frequent journeys to the nearest charging point. The extra cost of travel is effectively part of your fuel cost.
- You cannot spread the cost of large winter bills over the whole year if you pay for your fuel in advance week by week. A payment plan might be preferable if you could not afford to pay for your heaviest weeks' consumption from your weekly income.
- 'Self-disconnection' is a problem if you cannot afford to top up your meter and you may face intermittent or extended periods of disconnection. Fuel costs may take up too high a proportion of your income, particularly if you live in a property that is hard to heat or if your income is low.
- Paying back arrears and emergency credit can result in hardship. If a meter is set to collect arrears, a supply of fuel may not be available until the arrears charge has been paid. With some types of meters if you are away from home or cannot afford to charge the meter for a week, you have to insert two weeks' arrears before you can obtain a supply. With most types of prepayment meter, if you have used your emergency credit, you also have to pay the amount of the emergency credit before obtaining a supply. In some situations (eg, if you come out of hospital), you may be able to persuade the supplier to reset your meter – check first that this will not involve any extra cost.
- Your repayment of arrears may be highest when you can least afford it if you use the crude mechanical gas prepayment meters. These operate by overcharging for each unit of gas used, so the more gas you use, the more you

pay towards your arrears. This means there is no problem if you are absent from your home for any period of time – you will always get gas for every top up.

- Using the meters can be difficult, particularly if you have visual problems or disabilities. Note, however, that meters can often be re-sited free of charge to make them easier to use.
- Obtaining top-ups may present problems. Frequent journeys to buy them may present particular difficulties if you are caring for small children, are disabled or have limited mobility, or are in full-time work. It may be difficult to obtain top-ups outside shopping hours. Vending machines have had problems with jamming, vandalism and becoming full. Consider also the safety aspects of trying to obtain cards out of hours. Be sure to keep your receipts when charging keys/cards so that you have a record of payments.
- Keys can be easily lost or mislaid. If you lose your key, ask the supplier to replace it.
- You may be denied the option of changing to Fuel Direct (see p55) to pay your arrears if you already have a prepayment meter that has been reset to recover arrears.
- See p133 for the problems associated with illegally cloned keys.

Smart meters

Under a European Union Directive,[7] the UK is required to ensure that gas and electricity customers are provided with accurate meters which 'provide information on actual time of use' as far as is technically feasible. The Directive became law through the Energy Act 2008 which set the mandate for installing smart meters.[8] The government's aim is for every household to have a smart meter by 2020.[9] The government has placed responsibility on energy suppliers to replace over 53 million gas and electricity meters. A Code of Practice for the installation of smart meters has been issued.[10]

It is hoped that the introduction of smart meters will help Britain attain a low-carbon economy and help meet some of the longer-term challenges in ensuring an affordable, secure and sustainable energy supply.

The term 'smart meter' relates to the services and benefits obtained from such a meter rather than a specific type of technology. Smart meters measure your exact fuel usage and send the information electronically to your supplier without the need for meter readings to be taken.

Energy suppliers are developing smart pay-as-you-go meters alongside the smart meter rollout. Smart meters are expected to improve the experience of prepayment meter customers. Potential consumer benefits include:

- more convenience and choice in payment top-up methods;
- greater flexibility in friendly and emergency credit arrangements; *and*
- the ability to switch remotely between credit and prepayment modes.

Smart meters give better budget management tools such as low credit alerts and high consumption alerts which may safeguard against self-disconnection.

Advantages of smart meters

- There is no need for estimated bills, with suppliers able to read meters remotely via two-way communications technology. This should dramatically reduce the number of inaccurate bills issued – currently the biggest source of consumer complaints about energy companies.
- You can better judge your consumption of energy. You have real time information to help control and manage your energy use, save money and reduce emissions. You can compare the amount of electricity and gas you use today against what you used the day before, the week before, the month before and even the year before.
- It heralds the end of inaccessible meters under the stairs and in cupboards. The information on your energy use can be provided through a display device, via the internet or even through a mobile phone.
- Smart meters provide more detailed, user-friendly information on your energy consumption. The in home display shows you how much gas and electricity you are using and roughly how much it is costing you in pounds and pence. It is hoped that by knowing how much you are using, and how much your appliances cost to run, you may be able to reduce your energy consumption and save money.
- If you have a prepayment meter, it should be easier to top-up your meter. Cash payment is always be accepted, but suppliers can also offer more convenient ways to top up, such as over the phone, internet or with a special mobile phone app.

Disadvantages of smart meters

- A major disadvantage is remote disconnection of your gas or electricity supply in the event of non-payment, or an error creating an assumption of non-payment. This removes the protection afforded by most suppliers with the previous system – that suppliers could not disconnect your supply in the first instance without being given access to your property to perform the disconnection, or to fit a prepayment meter. With smart meters, entry is not necessary for disconnection. Ofgem has rules to ensure that suppliers treat disconnection as a last resort. Suppliers must ensure that vulnerable households are not disconnected. If you have an energy debt, discuss repayment options with your supplier.
- Suppliers are able to remotely switch the meter between credit mode and prepayment mode and do not have to physically visit the property to change meters. Energy UK guidance states that suppliers must first ensure that it is safe and reasonably practical for a customer to use a prepayment meter.[11]

- Concerns have been raised about purported health issues associated with the wireless technology used by smart meters to transmit information between supplier and meter.[12]

3. Payment methods

Your choice of payment method depends on the type of meter you have. You may wish to change your meter to allow you to use a particular payment method. If your circumstances change, you may be able to change to a more suitable payment method.

Each gas and electricity supplier publishes a code of practice on the payment of bills, outlining the various options available. Many also publish additional detailed information about the costs of the different options.

Credit meter payment methods

Payment can be made on receipt of a **quarterly bill** – either in cash at a customer service centre, by posting a cheque or giro, using a credit or debit card or paying directly into the supplier's account at a post office, PayPoint or through internet banking. A service charge may be applied if you pay at the post office. Using a credit card could be a very expensive way of obtaining credit because of the interest charged on balances. If you are in financial difficulty, consider the other budget options available.

Quarterly billing is currently the most expensive way to pay for your gas and electricity. Switching to direct debit will save you money. Furthermore, a succession of estimated bills can result in inaccurate billing, with you either paying too much or not enough. This can lead to a debt accumulating.

Direct debit is usually the cheapest way to pay for your gas and electricity. Your estimated annual costs are spread over 12 monthly payments (or four quarterly payments), which are deducted direct from your bank account. Payments should be enough to cover your annual consumption. If the direct debit payments are set too low you will accumulate arrears. Suppliers are able to adjust the direct debit amount but must inform you when this is the case.

Standing order is similar to direct debit with monthly or quarterly payments deducted direct from your bank account. The main difference is that you have to instruct the bank of any changes to the debited amount. Remember to take into consideration the extra costs of becoming overdrawn if you are considering standing orders or direct debits.

Setting up an **online account** with your fuel supplier and paying by direct debit attracts a discount and is usually the cheapest way to pay for your energy. This 'paperless billing' process enables you to view your bills online and submit regular meter readings.

There is a range of **budget schemes** which allow you to pay for your fuel on a weekly, fortnightly or monthly basis. This gives flexibility and helps you to budget, but is not the cheapest payment method. This is a useful way to pay if you do not have a bank account.

A **flexible payment scheme** enables you to pay any amount at any time at customer service centres, at a bank or by post. The amount paid is credited towards your next bill, which must then be settled each quarter. This is useful if you have a variable income.

PayPoint is a free national bill payment network aimed at households who prefer to pay utility bills in cash on a weekly, fortnightly, monthly or quarterly basis. Locations of PayPoint outlets can be found at www.paypoint.co.uk/paypointlocator.

Internet banking allows you to make payments to your fuel account from most bank accounts.

Prepayment meter payment methods

- **Key meters.** You are provided with a rechargeable 'key' when the meter is installed. The key can only be used in your meter. You need to charge the key by paying at a charging point (eg, at a PayPoint outlet) or you may be able to top-up at home using the internet and a device provided by your supplier. Your key is electronically encoded at the charging point with the amount you have paid. When the key is inserted into the meter, the amount of fuel you have bought is registered and the key is cancelled. A certain amount of emergency credit is usually available on these meters. Your key may be able to read your meter and pass on the reading when you charge it. Key meters do not need to be manually updated after a price rise.

- **Token and card meters.** Electronically coded payment cards or tokens, usually available in units of £5, can be purchased from local shops or customer service centres. Your account is credited every time you purchase a token/card. When the token or card is inserted into the meter it records the amount of fuel purchased and then automatically cancels the token/card. Token meters need to be manually adjusted after every price rise and you are at risk of continuing to buy top-ups unaware that you are no longer paying the correct cost for fuel. If there is a substantial delay in recalibrating the meter to reflect the price increase, a large debt may accumulate on your account. Your supplier should give you advance notice of a price increase and should take steps to recalibrate your meter as soon as possible.

Note: token/card meters have almost all been phased out and replaced with newer key prepayment meters.

4. **Fuel Direct**

Fuel Direct, also known as the Department for Work and Pension's 'third party deduction system', allows an amount to be deducted from your benefit entitlement and paid directly to your energy supplier. Payment for fuel arrears plus ongoing consumption, or for fuel arrears only (if you have a prepayment meter), is deducted directly from benefit. Deductions are made on a weekly basis from benefit payments, but are only paid to the supplier once every 13 weeks.

See Chapter 11 for full details.

5. **Choosing how to pay**

The legal provisions affecting your choice are broadly similar for both gas and electricity. An assessment is made of your creditworthiness and of your ability to pay for fuel in particular. If you cannot show creditworthiness in general, provisions relating to the requirement for security come into play. If you cannot show an ability to manage your gas/electricity account, both provisions relating to security and policies for debt management come into play. **Note:** an existing fuel debt is no longer an automatic bar to changing to a supplier who may offer a different type of payment method (see p56).

If you can show that a request for any kind of security from you is unreasonable (ie, you can prove your identity, show that you are creditworthy and have a good record of paying your gas/electricity bills on time, live in settled accommodation and are not in arrears with your bill), there is no reason for a supplier to attempt to restrict your choice of meter or method of payment. You should be allowed to pay using the method of your choice.

A supplier can restrict the choices available to you if it can demonstrate that it is reasonable to regard you as not creditworthy, even though you may not necessarily be in arrears with your gas/electricity bill. If it is reasonable for the supplier to require some form of security from you, your choices are restricted so that you are not allowed to accrue charges in the same way as customers paying quarterly in arrears. Payment to the supplier is required at least monthly, and possibly fortnightly or weekly, under the terms of the various budget schemes available.

If a supplier unreasonably requires some form of security, reference should be made to its code of practice. In the event of an ongoing dispute, contact Citizens Advice consumer service.

Payment in advance of receiving a supply, or prepayment, is at the other end of the scale. Suppliers claim it is costly for them, but actually it is the consumer who pays. This method of payment is the ultimate in security for the suppliers. No money – no fuel.

A prepayment meter may be your only option if the supplier has established a right to disconnect your supply, if you are in arrears and you cannot manage a payment plan. Similarly, if the supplier can demonstrate that a requirement for security is reasonable, and you cannot pay a deposit or arrange an alternative form of security, such as direct debit, a prepayment meter is the only alternative to disconnection.

Supplier discretion in switching with a debt

There is nothing to prevent a supplier allowing you to switch with debts in excess of £200. However, you may be required to pay what you owe before you switch. The decision to take on a customer with an existing debt is at the discretion of the company, and a number of suppliers appear willing to take on consumers who may have been previous customers and who have had good payment records. Switching supplier may also be a method of preventing disconnection from another supplier to whom a debt may be owed.

If you have a debt of up to £500 *and* use a prepayment meter, you may be able to switch to another supplier, taking the debt with you (see p105).[13]

Legal provisions: gas

The main provisions affecting your choice of meter or method of payment for gas are:
- condition 27.5–11 of the Standard Licence Conditions (SLCs) – this sets out the steps to be taken by a supplier when a bill is unpaid;
- SLC 27.3–4 – this enables gas suppliers to require reasonable security from customers.

Prepayment meters and credit meters have been put very much on an equal footing by the Gas Act 1995 and the way in which the suppliers' licences have been drafted. Under the terms of its original authorisation, British Gas was required to offer customers in arrears a payment plan as a way of repaying arrears. A prepayment meter could only be offered if this failed. This is no longer the case. Under the terms of the suppliers' licence, prepayment meters are no longer seen as a last resort option, but simply as an alternative method of payment.

Legal provisions: electricity

The main provisions affecting your choice of meter or method of payment for electricity as described above are:
- SLC 27.5–11 – this sets out the steps to be taken by a supplier when a bill is unpaid;
- SLC 27.3–4 – provides for security deposits, although no security deposit is needed for prepayment meters;

- your supplier's code of practice, approved by Ofgem, on the payment of bills – this sets out your supplier's policy on the treatment of customers in arrears and for requesting security deposits;
- section 20 of the Electricity Act 1989 – this allows distributors to request reasonable security (although there is no express power for contract suppliers to request reasonable security, it is clear from their licence conditions that they have such a power).

Notes

2. Standing charges and tariffs
 1 Condition 25A SLC
 2 Regs 8 and 16(2) WHD(MA) Regs

2. Types of meter
 3 Sch 6 para 1 EA 1989; Sch 2B para 6 GA 1986
 4 http://conversation.which.co.uk/energy-home/economy-7-energy-meter-clock-overpayments/
 5 Sch 7 EA 1989; SLCs
 6 Ofgem, *Domestic Suppliers' Social Obligations: 2014 Annual Report*, 8 September 2015
 7 2006/32/EU Energy Services Directive on Energy End Use and Energy Services
 8 ss88-91 Energy Act 2008
 9 DECC, *Smart Metering Implementation Programme, Programme update*, April 2012
 10 *Smart Meter Installation Code of Practice* (SMICoP), 27 June 2016, available at www.smicop.co.uk
 11 For suppliers' obligations, see www.energy-uk.org.uk/files/docs/Disconnection_policy/Sept15_EUK_Safety_Net.pdf
 12 Advisory Group on Non-ionising Radiation, *Health Effects from Radiofrequency Electromagnetic Fields – RCE 20*, April 2012

5. Choosing how to pay
 13 Ofgem debt assignment protocol for prepayment meter customers, www.ofgem.gov.uk/publications-and-updates/decision-make-modifications-gas-and-electricity-supply-licences-reform-switching-process-indebted-prepayment-meter-customers-debt-assignment-protocol

Chapter 5

· ·

Responsibility for the bill

This chapter covers:

1. Introduction

It is always worth checking whether you are legally responsible for a bill, particularly when you are in dispute with a gas or electricity supplier about arrears. It may be that you are not liable for all, or some, of the bill – perhaps because the bill was in the name of a partner who has left, a flat-sharer, your landlord or someone who has died. Electricity and gas suppliers may attempt to recover these sums from you – but they are not always entitled to do so.

Electricity

Liability for electricity charges is determined by the rules of the law of contract.[1] Generally, the person who signs a contract is the person who is liable to pay under that contract.

Some suppliers will accept you as a customer without you actually having signed a contract. A county court case covering the pre-1989 law suggests that it does not matter if you actually apply in writing or sign a contract, so long as it is clear who asked for the supply. While this case is helpful in establishing the liability to pay for customers who have requested a supply, whether in writing or not, it does not deal with the situation where nobody has requested a supply. Normally, once you move in you are responsible for any consumption, but only from the date that you moved in. For this reason, it is essential that you take a meter reading as soon as you move into a new property.

Electricity suppliers – or debt collecting companies to whom they assign the right to collect debts – may attempt to secure payment from people who have used the electricity supplied rather than chase the people actually legally liable. Previous suppliers may also assign the right to recover debts from new suppliers where the consumer has switched supplier. One example is where, on the breakdown of a relationship, if the bill was in the sole name of the partner who has left, the remaining partner is asked to pay the arrears.

Gas

The Gas Act 1995, as amended by the Utilities Act 2000, allows gas to be supplied by gas suppliers under the terms of contracts and 'deemed contracts' – instances where gas is supplied without any formal written contract being agreed. If the details of contracts are clearly confirmed in writing, disputes about liability are less likely to occur. However, the provision for deemed contracts (see p17) in many situations enables gas suppliers to hold owners or occupiers liable to pay a gas bill where they are the 'person supplied with gas'. If you are faced with this situation, consult Citizens Advice consumer service. Suppliers may be prepared to reach an agreement and settle an argument over liability by the part-payment of a bill in return for a new, signed contract being established.

---

2. When you are liable for an electricity bill

In this section, all references to 'the bill' refer only to 'charges due' for the supply of electricity and do not include other charges such as credit sales charges for appliances. See p114 for more information about 'charges due'.

Your liability for the bill depends on how, or even whether, you contacted the supplier to say you required a supply of electricity.

Your liability when you have signed for the supply

You are liable to pay an electricity bill to a contract supplier if you signed a contract.

Your liability for the bill begins from the date you stated you wanted the supply to start, providing your supply was actually connected on that date, or from the date you signed the contract.

You are solely liable for the bill if you alone signed for the supply, regardless of whether you live alone or with other adults. You are jointly responsible for the bill if you and one or more others also signed for the supply.

Your liability ends:[2]
- where you give at least two working days' written notice that you will no longer be an owner or occupier of the premises, on the day that you cease to be the owner or leave, as the case may be;
- where you did not give at least two working days' written notice that you were leaving, the earlier of any of the following three events:
  - two working days after you actually give written notice of ceasing to be an owner or occupier; *or*

– on the next date that the meter is due to be read; *or*
– when any subsequent occupier gives notice requiring a supply or signs a contract for a supply to the same premises.

You can also bring your liability to an end by terminating the contract in accordance with any provision for termination contained in the contract. You must give at least 28 days' notice of termination of a contract. You continue to be liable to the original supplier until another supplier takes over the supply to your home or until the supply is cut off altogether.[3]

Your liability when you contacted the supplier by telephone

Many electricity suppliers will connect your electricity supply without asking you to sign anything at all to confirm that you require a supply or that you accept liability to pay for any electricity supplied. Many operate a system where you negotiate with them by telephone to obtain your supply. In these situations, it is often clear exactly who is requesting the supply. Usually, payment is requested from the person who made the telephone call.

In the law of contract, in most situations, a verbal contract is as good as a written one, so what matters is not whether you signed a formal document but whether it can be shown that you were the person who asked for the supply and entered into the contract.

There is a danger that someone can contact the supplier over the phone to say that you want the supply to be in your name, but you have no knowledge of this and do not consent to it. In one case, a gas supplier attempted to obtain payment from a tenant. The tenant did not contact the supplier to request the supply in his own name and, in fact, paid his landlord for gas in with the rent. One bill was paid in the tenant's name, but there was no evidence that the tenant had made this payment. It was held that the tenant was not liable to pay for the gas consumed.[4] Although this case relates to gas, the principles apply equally to electricity.

Establishing joint liability for a supply may be problematic if you contact the supplier by phone. In practice, it is straightforward to ask the supplier to include someone else's name on the bill as well as your own, but what if that person denies s/he had an agreement to be jointly liable with you? It is best in these circumstances if everyone who requires the supply in her/his name signs for the supply. Your liability to pay starts from the date you ask for the supply to be put in your name. It will end on the basis of the rules given on p59. Always arrange a final reading of the meter, and take a reading yourself to check your final bill.

Your liability when no one has contacted the supplier

You may be liable to pay the supplier if:
• there is no tariff customer and no one has a contract; *or*

- the liability of the tariff customer or contract holder has come to an end; *and*
- you have, in practice, been supplied with electricity.

This situation often occurs when people move into new premises where the supply is already connected, or where the bill was in the name of another joint occupier but that person has left. There is nothing illegal about continuing to use the supply in another person's name, or where the bill is addressed to 'The Occupier', so long as you intend to pay for it. Anyone who uses fuel without intending to pay could be prosecuted for theft (see Chapter 9). The suppliers could rely on the law relating to 'unjust enrichment' to ensure payment in these circumstances. This applies when someone unjustly obtains benefit at someone else's expense. In England and Wales, the idea of unjust enrichment arises in the law of restitution; in Scotland, the equivalent is found in the law of recompense. However, if it can be shown that the supplier continued to provide fuel in spite of a request to disconnect or terminate a contract then the doctrine of unjust enrichment and the requirement of restitution do not apply.

Establishing who should pay the bill depends on the facts in each case.

Example

Sukie is a sole occupier and did not ask for a supply because it was already connected when she moved into her flat. She receives bills addressed to 'The Occupier'. Based simply on the facts, Sukie could be held liable for the bill from the date she moved in.

Where there is more than one occupier, establishing who is liable is often more difficult. Facts which might be relevant in establishing who is liable include:
- your status as an occupier;
- the extent of control you have over the use of fuel;
- your actual use of fuel;
- the degree of control you have over income within your household;
- the date you moved in;
- where the bill was previously in the name of another joint occupier, her/his status as an occupier and the date s/he left;
- whether you are registered at the dwelling for council tax purposes or for other utility bills.

It is advisable to give the supplier notice that you are leaving, to avoid disputes about the end of your period of liability, and supply a meter reading. Keep a record for yourself.

Your liability when someone else has been responsible for the supply

You are not liable to pay the supplier for the electricity bill if:

- you have not entered into a contract; *and*
- someone else is liable for the supply under contract and that person's liability for the supply has not come to an end (see p59).

You should also not be held liable for the supply when it is clear that someone else took responsibility for it, perhaps as a result of phone contact with the supplier. In one case, the court held that the wife of a deceased man should not be held liable for an electricity bill accrued by him in his name. She was only liable to pay for the supply following his death. The supplier had tried to argue that the woman should be held jointly liable for the debt, as she had benefited from the use of the supply – the 'beneficial user argument'. The judge declined to follow the county court case cited by the supplier in support of its argument.[5]

You are normally liable to pay for a supply of energy even if you did not read the contract before signing it. However, if there has been misrepresentation to induce you into signing, the contract may be set aside. Even if a set aside application is successful, you may still have to pay for the energy consumed if you have had the benefit of this.

Sending the bill to another person

If you are of pensionable age, disabled or chronically sick, the supplier may, where reasonably practical, send the bill to another person to deal with[6] – eg, a friend or relative.

Privity of contract

In some cases, a legal principle known as 'privity of contract' may be helpful in establishing liability. Basically, the principle states that only the two parties who have made and established the contract are subject to binding rights and obligations under it. If one party breaks the contract, the other party is the only person entitled to seek a remedy. Liability cannot be imposed on a third party who is a stranger to the contract unless an agreement or guarantee is given.

Examples
Ben is a student and lives in a house with other students. Ben's name is on the energy bill. An energy bill has gone unpaid. The supplier cannot approach Ben's parents and demand payment of the bill. Ben's parents are not liable for Ben's debts, since no contract was made with them.

Rhianna is a student and lives alone. Her father is the guarantor for her energy supply. A bill has gone unpaid. The supplier can approach Rhianna's father and demand payment.

Efah moves into a new property where previous occupiers have failed to pay an energy bill. Efah cannot be held liable for the bills left behind.

The doctrine of privity of contract is of importance where a house is in multiple occupation and residents are transient. In many cases, it is arguable that the long-term occupiers or the landlord are those who are liable under the supply contract, since these are the individuals who have actually reached a binding agreement with the supplier, either in writing, orally or by conduct.

3. **When you are liable for a gas bill**

The Gas Act 1995

Since 1 March 1996, gas has been supplied to existing customers of British Gas under the terms of a 'deemed contract'. From this date, new customers who request a supply from any supplier, including British Gas, are supplied under the terms of a contract. Any new customers who receive a supply of gas from a gas supplier without first entering into a contract are supplied under the terms of a deemed contract instead (see Chapter 2).

You are liable for a gas bill if:

- you were a tariff customer of British Gas on 31 January 1996. You will have had a deemed contract with British Gas from 1 March 1996. You are liable to pay for your supply under the terms of the deemed contract which applies to you;
- you entered into a contract with any supplier from 1 March 1996. You are liable to pay for your supply under the terms of the contract;
- you have a deemed contract with a supplier that started after 1 March 1996.

Former 'tariff customers' under the Gas Act 1986

If you were a tariff customer of British Gas immediately before 1 March 1996, you will have automatically become a customer with a deemed contract. You can be held liable to pay the bill under the terms of a deemed contract if you were the tariff customer under the provisions of the Gas Act 1986, before it was amended by the Gas Act 1995. This can only be decided under the terms of the law that applied at that time. A **'tariff customer'** was defined as 'a person supplied with gas' under the Gas Act 1986;[7] and should not be confused with customers under special tariff schemes run by individual fuel companies. Each supplier is required by the Gas Act to act in accordance with the Gas Code laid down in the Gas Act 1986.[8]

Termination of deemed contracts

The Gas Code does not specify when a deemed contract comes to an end. This means that a deemed contract continues until it is actively terminated. The contract starts from the moment that fuel is supplied to your home other than under a contract. The duration and methods for terminating a deemed contract should be specified, like the other terms, by each supplier. The deemed contract and liabilities under it continue in accordance with the terms of the deemed contract until it is actively brought to an end.[9]

Customers with contracts

Your liability for the bill

To be eligible to enter into a contract, you must be the owner or occupier of the property where the gas supply is needed. You must request a supply, though your request need not be in writing.[10]

A supplier may not enter into a contract with you if someone else has a contract for the supply of gas with another supplier for the same premises,[11] unless that contract will expire or terminate before you require a supply. In practice, you do not have to terminate your existing contract formally before entering a new one – if you sign up to a new contract, the changeover should be handled by the two suppliers.

You need to agree to the terms of your contract with your supplier. Terms of contracts are not individually negotiated. Your supplier has a 'scheme', approved by Ofgem, setting out the 'principal terms' of the various contracts on offer. You can request a copy of the principal terms, which must be sent within a reasonable period.[12]

Contracts may be for an indefinite period (known as a 'rolling contract') or a fixed period. Where a contract is due to come to an end, you have the option of agreeing either a new fixed-term contract or a new rolling contract with your existing supplier. If you do nothing, your supplier must switch you to the relevant cheapest rolling tariff it is able to offer. This switch happens automatically if no new contract is agreed.[13] The terms of contracts are restricted by provisions within a supplier's licence and regulated by Ofgem.

Ending your liability for the bill

You are liable to pay for the supply under the terms of the contract until one of the following applies.

- **Your contract comes to an end.** If a contract for a specified period is due to come to an end, the supplier must provide you with a statement of renewal terms no later than 42 days and no earlier than 49 days before the fixed-term period is due to end. This must be separate from any other document such as a bill, statement of account or marketing material.
- **You terminate your contract while you are still an owner/occupier of the premises.** A contract for a fixed period may be ended at any time during that

period if you give the supplier notice according to the contract and pay the supplier any termination fee referred to in the contract, unless either of the following conditions apply.

- **Your contract is of an indefinite length of time.** This also applies where an initial fixed-term contract has expired and you have continued to receive a supply from the company on a rolling contract.[14]
- **Your supplier has unilaterally varied a term in your contract.** The most obvious example is a price increase. However, any other changes made to your terms which significantly disadvantage you also apply here.[15]
- **You terminate your contract with your supplier and switch to another supplier.** See Chapter 2 for more information about this process. You do not have to pay a termination fee if you want to switch because your supplier has changed the terms of your contract unilaterally. You must notify your current supplier of your intention to switch on or before the date the variation takes effect in order to avoid the possibility of you being charged a fee. This notification does not have to be in writing. If you notify your supplier orally, make a note of the date and the name of the person you spoke to. You do not have to actually switch supplier or switch within a particular timeframe to avoid the termination fees, you simply need to notify your intention to do so.

 If you have arrears with your current supplier you can still in theory change supplier, but you must clear the arrears within 30 days.

 If you have a prepayment meter, your supplier can only block you from switching to another supplier if you owe more than £500.[16]

 Contact Citizens Advice consumer service if you want to terminate your liability and switch supplier and you believe that your current supplier is being obstructive.

- **You no longer occupy the premises.** The supplier must include a term that the supply contract ends two days after you have told the supplier of the date on which you stop owning or occupying the premises. If you do not give notice, the contract must end at either:
  - the end of the second working day after you have notified that you have stopped owning or occupying the premises; *or*
  - the date on which any other person begins to own or occupy the premises and takes a supply of gas at those premises.

 Where a contract is brought to an end in this way, you remain liable for the supply of gas to the property until the date on which that contract ends.

 If you give your supplier a minimum of two working days' notice, or if your supplier agrees to accept a shorter period of notice before you leave, your contract ends on the day you leave.

 If you do not give your supplier notice that you are leaving, your contract will not terminate, and you will continue to be liable to pay for the supply of gas until the earlier of:
  - two days after you have told the supplier you have left; *or*

– the date when another person requires a supply at the premises from either the same or a different supplier.

It is always in your interest to inform the supplier that you are leaving/have left. Make sure you arrange for the supplier to take a final meter reading and also read the meter yourself so you can check your final bill. Keep a copy of any correspondence.

- **The supplier varies the terms of your contract.** If your supplier increases the charges made under the terms of your contract or varies other terms of your contract unilaterally and the variations will significantly disadvantage you, the supplier must inform you in writing of the changes made. Note that Ofgem has ruled that suppliers must give you 30 days' notice of a price rise or any other change which will leave you worse off.[17] The supplier must also notify you that you have the option to change supplier, and direct you to a source of free and impartial advice about this process. If you decide to terminate your contract after receiving a price increase notice, and switch to another supplier, your old supplier must terminate your contract within 15 days of receiving confirmation that you now have a contract with your new supplier. Your old supplier cannot apply the price rise to your bill for the remainder of your contract with it.

- **The supplier's licence is revoked by Ofgem.** If this happens, your contract with that supplier is terminated. Ofgem will require another supplier to continue to supply you with gas under terms directed by Ofgem for up to six months.[18] You then become liable to pay the new supplier under those terms. If you choose to enter into a contract with the new supplier, the terms of the contract apply from the date you enter into the contract. The same applies if you choose to be supplied by a different supplier.

Customers with deemed contracts

Your liability for the bill

You are supplied with gas under the terms of a deemed contract if you are a consumer at the premises supplied with gas and:

- you are the owner or occupier of the premises supplied with gas, and the liability for the supply of the former customer (see p63) has come to an end or been terminated; *or*
- you became the new owner or occupier of the premises on or after 1 March 1996 and you have not entered into a contract with a supplier for a supply of gas; *or*
- your contract for the supply of gas has come to an end or been terminated and you continue to receive a supply of gas from the supplier; *or*
- you are being supplied with gas by another supplier because your supplier's licence has been revoked by Ofgem.[19]

The terms and conditions of a deemed contract are determined in accordance with a scheme set up by the supplier. They may include terms and conditions enabling the supplier to determine the amount of gas supplied to you if a meter reading has not been taken at the start of the deemed contract. Your liability is assessed under these terms until the earliest of the following three events:

- the date of the first meter reading; *or*
- the time the supplier ceases to supply you with gas; *or*
- the date you cease to take a supply of gas.

Disputes over the rate of gas consumption

In some cases, there may be a dispute over the amount of gas consumption, particularly where appliances have broken down or have not been used. In such a case, a supplier providing a gas supply under a deemed contract is required to act reasonably and take into account relevant consumption data for the premises.[20] For more on high bills, see Chapter 6.

Ending your liability for the bill

Your liability under the terms of a deemed contract continue until one of the following applies.

- **You enter into a contract while you are still an owner or occupier of the premises**. You continue to be liable under the terms of a deemed contract until your new contract takes effect.

  You may terminate a deemed contract at any time by giving the supplier seven days' notice. The notice period may be shorter if the supplier agrees. If you have not arranged to enter into a contract with the same supplier at the end of the notice period, your supply continues under the terms of a further deemed contract.

  If you intend to transfer to an alternative supplier, you must give your existing supplier at least 28 days' notice, unless the supplier agrees to accept a shorter notice period.

  You may not bring a deemed contract to an end without the agreement of the supplier, if you are being supplied by an alternative supplier because your supplier's licence was revoked by Ofgem. You can only bring such a deemed contract to an end:

  – with the agreement of the supplier (which will be given if you accept a contract with that supplier); *or*
  – by transferring to another supplier; *or*
  – by ceasing to take a supply of gas at the premises.

- **You cease to occupy the premises**. If you give your supplier a minimum of two working days' notice, or if your supplier agrees to accept a shorter period of notice before you leave, your deemed contract ends on the day you leave.

If you do not give your supplier notice that you are leaving, your deemed contract does not terminate and you continue to be liable to pay for the supply of gas until the earliest of:
- 28 days after you inform the supplier you have left;
- the next date the meter is due to be read;
- the date when another person requires a supply at the premises from either the same supplier or a different supplier.

4. Common problems

Your liability when your name is on the bill

Electricity

The person named on a bill is not always liable to pay. Sometimes only one person is actually named on a bill, even if several people signed the notice requiring a supply or the contract or no one signed anything. For example, suppliers sometimes ask outgoing occupiers the names of the next occupiers. You may find that your name is on a bill without your ever having had any contact with the supplier. A person who is named on the bill has sole liability for the bill if s/he alone gave written notice requiring a supply or s/he alone made a phone request for the supply. Otherwise, the name on the bill is only evidence of who might be liable.

Gas

If you have a deemed contract with British Gas (see p63), you need to check that you are liable under the provisions for deciding who becomes a customer with a deemed contract (see p66).

If you have entered into a contract with a gas supplier and your name is on the bill, you are liable under the terms of the contract (see p16).

Who is liable when no one is named on the bill

Electricity

When no one is named on the bill, liability depends on the facts.

Gas

If no one is named on the bill, you may be liable for the bill under the terms of a deemed contract (see p63).

Moving in: becoming liable for the supply

When moving to a new address, make a note of the meter reading and preferably agree the reading with the last occupier(s), so that you can use this as evidence of when your liability for supply started. If possible, arrange to have the meter read by the supplier, and inform it that you require a supply. Check your first bill carefully to ensure that the dates used by the supplier are correct and the bill does not include the previous occupiers' charges. You cannot be held responsible for the previous occupiers' arrears of electricity or gas.

Electricity

If an electricity supplier does not routinely use application forms, consider whether you should request a supply in writing (see p29). One person can give notice if that person wants to take responsibility. If you are a joint occupier who wants to share liability, ensure that everyone living in the property signs the application form, letter or contract.

If one joint occupier moves out and you move in to take her/his place, there is nothing to prevent you from giving notice specifying that you are replacing a joint occupier or arranging a new contract. Liability depends on who signs the new arrangement.

Gas

If you do not inform a supplier that you have moved in, the last supplier to supply gas at the premises is entitled to charge you for any gas you have used under the terms of a deemed contract. To avoid this happening, take proactive steps to contact the supplier to obtain a supply. You are initially liable under the terms of a deemed contract, until you enter directly into a contract with the supplier.

Moving out: ending liability for the supply

If you are an occupier and you give a supplier proper notice that you are leaving, you should not be held liable for the fuel used after you have left.

You remain liable for six years after the date on which a bill for gas or electricity fell due in England and Wales; five years in Scotland. Thus, if you change supplier but have an outstanding debt to a previous supplier, the previous supplier has a right to bring legal action against you for up to six years. Thereafter, the debt becomes irrecoverable under the Limitation Act 1980. In Scotland, the limitation period is five years.[21]

Electricity

In the absence of notice, ending liability depends on the terms of your contract with the supplier.

Gas

There are provisions in both the Gas Act 1986, as amended by the Gas Act 1995, and in suppliers' licence conditions that set out when your liability for gas ends under contracts (see p64) and deemed contracts (see p67).

Who is liable when the person named on the bill has left

Sole liability

Electricity

A person who is named on the bill has sole liability for the bill if s/he alone became the customer by giving written notice requiring a supply or entered into a contract. Her/his liability ends either when s/he left if s/he gave notice of leaving or with the passing of time. As a joint occupier, spouse or co-habitee, you cannot be held liable for her/his bill if you have not given notice requiring a supply or entered into a contract. If you remain in the property after the co-habitee has left, you should request a new supply in your name. This is the point at which your liability for supply commences (see p62).

Gas

If you alone entered into a contract for the supply of gas, your responsibility for that supply ends either on your terminating the contract (see p64) or with the passing of time (see p67). One or more remaining occupiers may subsequently have responsibility for the supply under the terms of a deemed contract (see p66).

If you are liable under the terms of a deemed contract, your liability ends either with the termination of that deemed contract or with the passing of time (see p67). The liability of the remaining occupiers is also under the terms of a deemed contract.

Shared liability

Electricity

If the person who left gave written notice or entered into a contract to obtain the electricity supply, her/his liability ends either when s/he informs the supplier s/he is leaving or with the passing of time (see p59). Any remaining occupiers who originally gave notice in writing or entered into the contract are liable for the arrears along with the person who has left.

Where nobody gave notice or signed a contract, liability for the arrears depends on the facts. You could still be held liable for all of the arrears, but may be able to negotiate a compromise with the supplier (look at the supplier's code of practice on payment of bills, which may contain an indication of the supplier's attitude).

As a joint occupier or sharer, you could ask for the amount of the arrears to be apportioned between the people responsible for the bill, particularly if the supplier knows the whereabouts of all the parties. Electricity suppliers are entitled to refuse to supply an occupier who owes arrears[22] and, in any event, could pursue each debtor separately through the courts.[23] If the occupier who left was your partner, and you had little or no control over the income of the household (eg, only your partner had a wage or received benefits), you could argue that you should not be held responsible for any arrears that accrued while your partner was present and ask the supplier to pursue your partner for the arrears.

Gas

If you were jointly supplied under the terms of a contract, the terms of the contract apply. There may be scope for you to argue that any arrears should be apportioned between all the parties to the contract. When the previous joint occupier leaves, make sure you inform the supplier of the meter reading and/or apply for a new contract/deemed contract in your own name so the arrears relating to the joint occupancy are clearly established.

Where a joint deemed contract comes to an end because one of the occupiers has left, inform the supplier and establish any arrears relating to the period of joint occupancy.

The strict application of this legal position on liability for a bill can create problems for many customers with children left with large arrears after a spouse/partner has left. Gas suppliers should, therefore, be urged to treat these situations sympathetically and positively, taking into account the individual circumstances of each case. They should also take into consideration the existence of formal or informal agreements between the parties concerned for responsibility for household expenses, including gas. Seek advice from Citizens Advice consumer service if you do not feel your gas supplier is acting reasonably.

Who is liable when the person named on the bill dies

When the person named on the bill dies, the supplier may attempt to secure payment from someone who was living with her/him. In these circumstances, the situation is as outlined above. Where no one else is liable for the bill, any bills outstanding can be charged to the deceased's estate. This means that outstanding bills must be paid for out of money belonging to the deceased, or out of the proceeds of the sale of any belongings. The cost of the funeral and any costs involved in dealing with the estate take priority over all other debts except 'realised securities' such as a mortgaged property. So, if a consumer dies with an outstanding mortgage and the house is sold to meet this debt, anything left will go first to pay for the funeral and administration costs; any outstanding fuel bills are a lower priority. See CPAG's *Debt Advice Handbook* for more on the priority of debts.

If the person left nothing, the bill lapses and the supplier must bear the loss. If you have paid the bill of a person who has died, in the mistaken belief that you were responsible for doing so, the supplier can usually be persuaded either to credit your own account or refund the money. If the supplier refuses to do so, seek legal advice about the options available to you.

Assignment of outstanding charges to your new supplier

You might become liable for an old energy bill where the supplier has assigned the outstanding debt to your new supplier. If you failed to pay the old supplier within 28 days of the charges being due, it may assign the debt to your new supplier. If your new supplier agrees, it can take over the debt where:

- it has become due to the first supplier;
- it had been demanded in writing; *and*
- you were notified that the charges might be assigned.

However, it is more likely that the transfer will be blocked by a supplier within the framework laid down by Standard Licence Condition (SLC) 14.

Domestic customer transfer blocking

Under SLC 14, the supplier may prevent a proposed supplier transfer where:

- there are outstanding charges; *or*
- the transfer was started in error; *or*
- the terms of the contract prevent a transfer – eg, the contract does not end until a fixed date in the future.

A supply transfer cannot be prevented where gas is supplied by a prepayment meter and you have agreed to pay the existing charges. Nor can a transfer be stopped where the supplier has increased charges but has not re-set the prepayment meter within a reasonable period, and the outstanding charges only relate to the period since a price increase.

The Limitation Act 1980

Energy companies may sometimes attempt to pursue debts more than six years after the sum fell due. A six-year recovery limit is placed on sums due under contract (actual or deemed) by the Limitation Act 1980. In Scotland, the limitation period is five years.[24] The Act applies to a contract to supply energy in the same way that it applies to other contracts. This means that if you have not acknowledged the debt for six years then the energy company cannot take court action to recover the money claimed. Making a payment to the energy company for the sum claimed would constitute acknowledgment, as would any written correspondence about the debt. It is not usually possible to acknowledge a debt verbally – eg, over the telephone. You should exercise caution about your contact with the energy company if the debt is nearing the six year limit, as acknowledging the debt can start the clock running again.

Do not ignore any summons from the county court and seek advice about whether you need to defend the claim – eg, because the time limit for taking action has expired. If you do not do this within 14 days of being sent the proceedings a default judgment may be issued automatically against you even if there is no valid basis for the claim.

If the supplier has obtained a county court judgment within the relevant limitation period, then in theory there is no limit on the amount of time that the energy company can pursue you for the balance due under the judgement.

However, if the judgment is over six years old the creditor may need to obtain the permission of the court to enforce the debt. The court will consider whether the supplier has acted reasonably and promptly in bringing any claim against you. For more information on court proceedings, see Chapter 14.

Harassment by suppliers and debt collectors

In some cases, a supplier may wrongly attempt to pursue a claim against you long after any supply and any liability has ended. Such acts may constitute harassment if the supplier persists despite you establishing the true position.

In extreme cases, sending demands for payment accompanied by threats of disconnection or referral to credit reference agencies may constitute harassment both in civil and criminal law.[25]

In other cases, an old debt may be assigned by the energy company to a firm of debt collectors. These companies may contact you frequently by letter and by telephone, threatening to take legal action against you. Often they are based far from where you live and have no intention of issuing any sort of legal proceedings or visiting you, despite claims in the letter that they will do so (see Chapter 14).

If you deal with these companies, do so in writing rather than by telephone, bearing in mind the advice above about the limitation period if you have not had any contact with the supplier or its representatives for some time. You should seek advice if possible before responding.

Any such claim should be closely examined to ensure that there has not been a mistake. If contacting a debt collecting company, request copies of all the alleged paperwork on which the claim is based. As fuel debts are based upon contract, there should be evidence of a valid assignment of the debt between the energy company and the debt collector. An **'assignment'** is a legal document whereby the legal right to claim the debt is transferred from one person to another, including the right of enforcement. The details of any assignment must match precisely the amount being claimed. You should ask the debt collecting company to produce a copy of any assignment from the original supplier.

Debt collectors have no right of entry to your home. If they make threats or commit acts of harassment contact the Financial Conduct Authority (FCA) who assumed responsibility for regulating such companies from 1 April 2014. The FCA's *Consumer Credit Source Book* may be useful here; particularly section 7 which refers to the conduct of external debt collection companies. Although gas and electricity contracts are not regulated agreements, the guidance set down in section 7 can be used as useful guidance to what constitutes reasonable behaviour. See www.handbook.fca.org.uk/handbook/conc/.

Notes

1. Introduction
1 Condition 22.1 SLC

2. When you are liable for an electricity bill
2 Condition 24 SLC
3 Condition 24 SLC
4 *British Gas plc v Mitchell* (unreported) Pontefract County Court, May 1994
5 *Faulkner v Yorkshire Electricity Group plc* [1994] Legal Action, February 1995, p23
6 Condition 26.1(b) SLC

2. When you are liable for a gas bill
7 s14(5) GA 1986, repealed by GA 1995;
8 Gas Act 1986 schedule 2B
9 *Laverty and others v British Gas Trading Ltd* [2014] EWHC 2721
10 Condition 22.2 SLC
11 Condition 22 SLC
12 Condition 22 SLC
13 Condition 22 SLC
14 Condition 24.3 SLC
15 Condition 24.3 SLC
16 Ofgem debt assignment protocol for prepayment meter customers letter, 22 September 2014 (https://www.ofgem.gov.uk/sites/default/files/docs/2014/09/open_letter.pdf)
17 Condition 23.4.a SLC
18 Condition 8.2(b) SLC
19 Conditions 7 and 22 SLC set out the obligations of a supplier under deemed contracts
20 Condition 7.9 SLC

4. Common problems
21 Prescription and Limitation (Scotland) Act 1973
22 Condition 22 SLC
23 *Laverty and others v British Gas Trading Ltd* [2014] EWHC 2721
24 Prescription and Limitation (Scotland) Act 1973
25 *Ferguson v British Gas Trading Ltd* [2009] EWCA Civ 46

Chapter 6

....

High bills

This chapter covers:
1. Amount of the bill (below)
2. Accuracy of the bill (p77)
3. Accuracy of the meter (p81)
4. Meter faults and faulty appliances (p85)
5. Overcharging by prepayment meters (p87)

_____

1. **Amount of the bill**

This chapter looks at ways of checking whether you are paying the correct amount for your gas and electricity. If your bills are correct, see Chapter 4 for how to pay; chapters 11 and 12 for financial and other help with paying them; and chapters 7 and 8 if you are in arrears and/or facing disconnection. Chapter 14 suggests remedies for when a supplier charges you the wrong amount.

....

Understanding your bill

Your bill must show what you have paid and what you owe, a summary of your energy usage and details of your supplier's cheapest deal. Ofgem's factsheet 124 may be useful.[1]

Suppliers must provide a bill or statement of account at least twice a year, or quarterly if you request a quarterly bill.[2] All bills should be 'in plain and intelligible language'[3] and free of charge.[4] A supplier cannot charge for providing you with a bill, details of consumption used to calculate the bill or any statement of account. However, your supplier can charge for providing copies of bills that have already been sent to you.

See Appendix 3 for help understanding fuel bills.

....

Duty on suppliers to obtain meter readings

Suppliers are under an obligation to take all reasonable steps to obtain accurate meter readings from customers annually.[5] This does not apply to prepayment meter customers.

The supplier may fulfil its obligation by accepting a meter reading which you take and supply to the company, which should then be reflected in your next bill.[6] If the supplier does not accept your reading, it is under an obligation to take all reasonable steps to contact you to obtain a further reading[7] or it must obtain its own reading.

Check consumption

If a fuel bill seems to be too high, first check whether your consumption has actually increased. You can do this by comparing the units consumed with those used during the same period in previous years, by looking at bills for previous years. Standard Licence Condition (SLC) 31(A) makes this process relatively straightforward as it requires that your bill should include a comparison of your fuel consumption for the period covered by the bill with your consumption for the same period in the previous year. It should also include, among other requirements:
- the exact tariff name;
- your annual consumption details;
- your estimated annual costs.

If you do not receive regular bills (eg, because you have a prepayment meter), your supplier must send you an annual statement which contains the same information.

If you have not kept previous bills, ask the supplier for copies. Some suppliers may charge for these. If the charge seems unreasonably high, contact Citizens Advice consumer service for guidance.

Bear in mind that consumption will fluctuate seasonally, so you are likely to consume more fuel in cold winter months than you are in the summer.

Suppliers must give you 30 days' notice of a price rise or any other change to billing which will leave you worse off.[8] Suppliers also now have to notify you of how price increases will impact, in pounds and pence terms. SLC 31(A) requires that suppliers include on your bill the exact text: 'Remember – it might be worth thinking about switching your tariff or supplier'.

Your supplier must also include information about sources of independent and impartial advice and information about changing your supplier and provide a website link to energy efficiency guidance.

Exceptional reasons for a high bill

Check whether consumption is higher than usual because of exceptional reasons – eg, because of:
- an emergency, such as a flood;
- a new heating system that you are not used to operating or that might be defective;
- a period of exceptionally cold winter weather;
- someone in the household being ill, leading to higher heating costs.

If you think the bill is accurate after taking any exceptional circumstances into account, see chapters 4, 11 and 12 for help on how to pay.

Other charges in the bill

Bills may also be high because they contain items other than the cost of fuel and standing charges. The supplier may also include charges for disconnection or reconnection, and for replacing meters. Check to see if these have been lawfully charged by reading Chapter 3 on supply and charges for connecting supply, and Chapter 8 on disconnection for arrears.

2. **Accuracy of the bill**

Always check that the supplier has calculated your bill correctly. If the units consumed are correctly recorded, check that the cost has been correctly calculated.

If disconnection is threatened but you think the bill is inaccurate, tell the supplier immediately. A supplier cannot disconnect in respect of part of a bill that is genuinely in dispute.[9] However, you cannot avoid paying for gas or electricity used simply on the basis of a mistake resulting in an inaccurate bill – the supplier has the legal right to make you pay for what you have used, even if it initially made an error on the bill.[10] In particular, you should take steps to pay for that part of the electricity bill that you do not dispute, or you could be disconnected for that alone. If you cannot afford to pay the amount you agree you owe in full, set up a payment plan to demonstrate that you are willing to pay for what you have used (see Chapter 7).

If you experience difficulty in dealing with a supplier, particularly if staff at call centres are unable to help, be prepared to lodge a complaint. This will usually result in reaching a more senior member of staff to deal with the issue.

In addition to the statutory requirements set out in Standard Licence Condition (SLC) (31A) which have raised the bar considerably in terms of the amount of information suppliers must provide on bills, most suppliers[11] are also committed to the Code of Practice for Accurate Bills,[12] which sets out minimum standards you can expect from your supplier. These include:
- bills and statements that are clear, accurate, useful and on time;
- advice on checking energy use;
- obtaining and recording the most up to date and accurate meter reading;
- a commitment to the back billing rules (see below).

The clause relating to back billing within Energy UK's Code of Practice for Accurate Bills applies to domestic credit customers, as defined in the supplier licence agreement. It does not apply to customers whose express terms and conditions exclude them from receiving bills or statements.

Although prepayment customers are not explicitly covered by this clause, the code states that suppliers will apply similar principles in relation to debt over 12 months old.

Under these 'back billing rules', where your supplier has, through its own error, failed to issue bills, it will not bill you for any amount which dates back for longer than a year. You must be able to show that you have not avoided payment – eg, by refusing to cooperate with attempts to read the meter or to resolve queries. You must also be able to show that you have made a reasonable attempt to contact a supplier to make or arrange payment either in full or by way of an instalment. Every back billing case should be assessed on an individual basis.

Where a supplier issues a bill which has these principles applied, it will credit your account with the value of the unbilled energy consumed over 12 months ago, taking into consideration any payments already made by you or credits applied to the account, so that you are not required to pay any additional sums towards this previously unbilled energy consumption.

Billing delays

You may find that the supplier has billed you for a longer period of time than normal and, when you finally receive the bill, it covers that whole period. Suppliers are not obliged to bill you at any particular interval, although they will normally send bills every three months or six months. If you have a prepayment meter, statements are usually sent to you at least once a year. If regular bills have not been sent, or are late, point out to the supplier that the high level of debt is partly its fault by making it difficult for you to monitor or modify consumption. It should be possible to negotiate time to pay.

Bills sent after a long time should be examined with care: the charges may have changed since the previous bill. If the prices have gone up or you have changed your tariff, check to ensure that the supplier has charged accurately and at the correct unit rate. If it has not, the bill should be reduced appropriately. Also check that the supplier has complied with the requirement at SLC 23 to give you 30 days' notice of any changes to the cost of your supply.

There have been cases of bills being sent out late by months or even years. Unless fraud is involved (see Chapter 9), charges cannot be recovered:

- in England and Wales, if you last made a payment or acknowledged the debt more than six years ago;[13] *or*
- in Scotland, if the supplier has not raised legal proceedings against you for the charges and you have not acknowledged the charges for five years.[14]

Otherwise, you must pay for the gas or electricity that was supplied to you (see Chapter 5). With a bill that is very late, much of the bill may be estimated (see p79). The longer the delay, the better the chances of having part or even the entire bill written off, or being given time to pay. Be prepared to negotiate and place proposals in writing, setting out your circumstances as necessary. Citizens Advice consumer service can advise and the Energy Ombudsman could apply pressure on the supplier to settle on a reasonable solution as appropriate (see Chapter 14).

Obtaining information under the Data Protection Act

Your supplier is likely to hold a large amount of information about you electronically. This may include details of previous bills and consumption, credit ratings, and prosecutions. If a supplier refuses to provide information voluntarily, you may still be able to get hold of this information by using your rights under the Data Protection Act.

To exercise your rights, contact the supplier in writing by email or post. The Information Commissioner's Office (ICO) website contains a useful guide to doing this and a sample letter.[15] You have to pay a fee of up to £10. The supplier will then provide the information.

As well as the right to be supplied with copies of information, you have the right to:
- have inaccurate information corrected;
- claim compensation for loss caused by inaccurate information;
- complain to the ICO if the supplier fails to provide the information, to correct inaccuracies, or to obtain and process information fairly and lawfully.

Estimates

Check to see whether the bill has been estimated. Your supplier must clearly indicate on your bill if an estimated reading has been used. The abbreviation 'E' ('estimated') may be next to the figure the meter reading. It must also clearly indicate if the figure used is based on a meter reading provided by you, or if the supplier has read the meter. Any letter or symbols used should be explained on the bill.

If practical, take your own reading if the bill is based on an estimated reading. (see Appendix 2). Do this as soon as possible after receiving the estimated bill and make an allowance for units consumed since the date of the bill so that the comparison is as accurate as possible. You can calculate an average daily rate and deduct that amount for the appropriate number of days.

Alternatively, to decide if an estimate is unreasonable, compare the amount of fuel used over the same period in the previous year with the estimated consumption on the present bill. If a bill covering the same period is not available (eg, because you have recently moved), try making a reasonable estimate using any bill you do have, or by calculating how much would be used by the appliances you have and the frequency with which you use them.

If you do not think the estimate is accurate, notify the supplier. Call the number on the bill, or record your own meter reading on the back of the bill and ask for a more accurate bill to be sent later. If you are able to, pay the amount claimed on the bill and the supplier should work out the difference it owes you. It is advisable to back up a telephone request by email, keeping a copy. If your own reading shows that the bill is an overestimate, the supplier must take all reasonable steps to reflect your meter reading in your new bill – in line with SLC 21B.

However, remember that if a bill has been underestimated, a higher bill will result from any complaint. It is still important to obtain an accurate bill from your supplier which reflects your actual usage so that you can start addressing the bill promptly. If you are worried that receiving a higher bill will cause you financial hardship, see Chapter 7. Also, be aware of cases where a high bill is the result of a low estimate on a past bill followed by an accurate meter reading later. This is one of the most common reasons for an extraordinarily high bill: several quarters' bills are significantly underestimated, followed by an actual meter reading by the supplier which brings the account back up to date. This results in a large bill, incorporating the usage not paid for in the previously low estimated bills. For this reason, it is crucial to give an accurate reading yourself whenever you receive an estimated bill. The points made in respect of billing delays on p78 also apply here.

Errors in reading the meter

Although actual meter readings are likely to be correct, even the supplier's own meter readers can sometimes make mistakes. You can use the same methods discussed above to detect errors, but the supplier might want to take a second reading to check any reading which you have taken. If the supplier has made a mistake, the bill should be amended.

Errors in assigning bills

In some instances – eg, where there are properties in multiple occupation or blocks of flats where meters are often grouped together – readings can sometimes be attributed to the wrong meter. Every gas and electricity meter has a unique reference number that effectively links the meter to a specific address. For electricity meters, this is known as the meter point administration number (MPAN). The format of MPAN is standard and consists of 21 digits. For gas, it is known as the meter point reference number (MPRN). The format of MPRN is also fairly standard and consists of between six and 10 digits. Occasionally, these are also referred to simply as 'meter numbers' or as 'M' or 'S' (supply) numbers. The reference number(s) must be shown on your fuel bill(s) and your annual statement. It may be worthwhile checking with your supplier that the MPAN/ MPRN number on your fuel bill 'belongs' to your address.

If you can't find your MPAN number, contact your Distribution Network Operator (DNO). For a full list of DNOs, see www2.nationalgrid.com/uk/Our-company/electricity/Distribution-Network-Operator-Companies/. For your MPRN number, call the M number enquiry line on 0870 608 1524.

Smart meters

The new generation of smart meters (see p51) allow your supplier to read your meter remotely, without the need to estimate bills or even to visit your home to take a manual meter reading. Smart meters will remove the need for bills to be estimated at all, allowing you to pay only for the fuel you actually use.

Fuel companies are responsible for meeting the cost of installing the new meters. A code of practice for the installation of smart meters is available at www.smicop.co.uk. The government anticipates that all homes will have access to smart meter technology by 2020, including customers using prepayment meters. When your smart meter is installed, you should be shown how it works and what information can be accessed. You will also be offered energy efficiency advice.

The Department for Business, Energy and Industrial Strategy anticipates that smart meters and tools such as the in-home display will offer you greater control over your fuel costs by allowing you to access detailed information about your usage levels, showing not only the amount of energy being used, but the cost of this energy in pounds and pence.

3. **Accuracy of the meter**

Even if the details on the bill appear to be correct, the meter itself may be faulty. Meters must be approved and certified by meter examiners.[16] If, as is the normal situation, the meter belongs to the supplier (or gas transporter), it is responsible for keeping it in proper working order.[17] An electricity meter is deemed to be accurate if it does not vary more than +2.5 per cent to -3.5 per cent from the correct reading. For gas meters the limits are +2 to -2 per cent.

Note also that if gas or electricity appliances are old and/or have not been serviced recently, they may no longer perform in accordance with their rating.

Prepayment meters

It is important when moving into a new property with a prepayment meter that you ensure the supply is in your name from the date that you started living at the property. Otherwise, you may find that what you pay for fuel is not credited to your account or that you are inadvertently paying for the fuel debt accrued by the last occupant.

Contact the fuel supplier at your new property as soon as you move in. Do not use the previous occupants' payment card/key as you may inadvertently make payments to their account.

Ask your supplier to provide you with a new card or key in your name. Some suppliers can reset a key or card remotely but will still require you to provide them with a meter reading from the date that you moved in. In other cases, your supplier may need to visit your property to reset the meter, depending on the age and type of device you have. Contact your supplier to establish their procedure for your meter.

To check whether a prepayment meter is registering correctly or to find out the rate of charge, see Appendix 2.

Checking your meter

Gas meters must be installed in a readily accessible position and, if situated in a box or compound, you must be given a key. If you are disabled, your meter may be repositioned or replaced with an adapted one for free[18] – ask your supplier to do this for you.

You can check your meter by using the following method.

- Switch off all appliances, including pilot lights.
- Read the meter (see Appendix 3).
- Turn on an appliance with a known rate of consumption and note the time.
- Leave the appliance on for a measured period of time, preferably in whole hours.
- Switch off the appliance and read the meter again.

If everything is working properly, the following formulae should work.

Electricity
Difference in readings = rating of appliance (kW) x time on (hours).

Example
If your electric fire has a one kilowatt (1kW) rating and is switched on for one hour the figures on each side of the '=' sign should both be '1'.

Where an electricity meter is faulty and its readings are outside the statutory margins of error, you fall within the standards regulations and may be entitled to bring a claim. For electricity, the margin of error is laid down by paragraph 13 of Schedule 7 of the Electricity Act 1989.

Gas
Difference in readings =

$$\text{(hundreds of cubic feet)} \times 3.6 = \frac{\text{appliance rating (kW)} \times \text{time on (hrs)}}{\text{(calorific value* for region)} \times 2.83}$$

*The amount of heat produced by burning a specific amount of gas

Example
The rating for an average gas fire is 2.5 kilowatts (kW), assuming that the fire is on full and all the elements are being used. If it is switched on for one hour in an area where gas has a

calorific value of 38.2, then the figures on each side of the '=' sign should both be 0.023 – ie, the reading should be 2.3 cubic feet.

To find out the rating of an appliance, if it is not marked on the appliance itself, contact the manufacturer, the appliance supplier or the gas supplier. The calorific value for the area will be shown on your gas bill, or you can ask your supplier.

Where a reading falls outside a specified margin of error, there may be the basis of a claim under the Gas Standards Regulations. The relevant standards are those set out in the Gas (Meters) Regulations 1983.

Standards of performance

The Electricity and Gas (Standards of Performance) (Suppliers) Regulations 2015 may apply where you or the supplier considers that a meter may be faulty.

Where the supplier needs to visit your property, it is expected, within a reasonable time, to offer you an appointment during working hours that is no more than four hours long.[19] The supplier must not rearrange an appointment with less than one working days' notice without your agreement.

Standard meters

Once notified, the supplier is normally required, within five working days, to:[20]
- complete an initial assessment;
- take an appropriate action; *and*
- offer to confirm in writing the nature and outcome of that initial assessment, the actions it will take and the timescale within which those actions will occur.

If it does not respond within the five days, you are entitled to a £30 payment, unless there is an exception (see below).[21]

Prepayment meters

Where you notify the supplier of a loss of supply of gas or electricity from the meter, the supplier must restore the supply within three hours on a working day; or four hours on any other day.[22] This may be done remotely or by visiting your property. Where you contact the supplier outside normal working hours, the communication is deemed to have been made the next working day. If the supplier does not meet these standards, you are entitled to a £30 payment, unless there is an exception (see below).[23]

Exceptions

Circumstances in which you will not be entitled to a £30 payment include:[24]
- that you requested the supplier not to attend the premises;
- that you requested that the supply not to take any action, or any further action;
- where the prepayment meter is found to be working correctly;

6

- it was not reasonably practicable for the supplier to meet the individual standard of performance as a result of not obtaining necessary access to the premises, severe weather conditions and other exceptional circumstances beyond the control of the supplier.

Meter examiners

If you think that a meter is not functioning properly, complain first to the supplier (or gas transporter, as appropriate) who can check it.

Your supplier may try a simple test of the meter without moving it, such as putting a check meter to run alongside your meter for a week or two or undertaking a meter test. If you are still not satisfied, refer the matter to a meter examiner. The supplier/transporter can also make the referral.

If a meter seems to be over-registering, you should, if you can afford to, pay the supplier for the amount of fuel you think you have definitely consumed, without waiting for the examiner's decision. This would help effectively neutralise the threat of disconnection and the dispute would be considered a genuine one (see p77).

If the meter is removed, a substitute meter should be installed. There should be no charge for this.

Electricity

The electricity meter examiner's service is contracted out to the Regulatory Delivery directorate, now part of the Department for Business, Energy and Industrial Strategy (DBEIS). It replaced the National Measurement and Regulation Office from March 2016. An examiner will test the meter and the supply at your home. The supplier will be invited to send a representative. The meter may then be removed for further tests. An electricity meter examiner's services are free but the supplier is likely to charge. Take a note of the reading on the meter before it is taken away.

If a notice is served by you, the electricity supplier or anyone else interested in the matter, then no one can alter or remove the meter until the dispute is resolved or an electricity meter examiner has finished her/his examination.

The findings of an electricity meter examiner can be produced in court and are presumed to be correct unless proven otherwise.

Gas

Gas meter examiners are also contracted out to Regulatory Delivery under DBEIS. If there is a dispute about meter accuracy, they would not come to your home but would request that your supplier sends them the meter for examination. Make a note of the reading before it is taken away.

The supplier may charge you for the removal of the disputed meter, installing a replacement, transporting the disputed meter for testing and reinstalling the meter at your property. This charge will be refunded if the disputed meter is found to be operating inaccurately. The charges made for meter examining vary – check with your supplier.

On completion of the examination, a gas meter examiner issues a test certificate with details of her/his decision. The decision is final and binding as to whether the meter is working properly or not.

Results

If your meter is found not to be working properly, the supplier has to make a refund or an extra charge to you. The amount of the refund or charge depends on how long and by how much the meter is thought to have been registering incorrectly.

For electricity, the meter examiner has a duty to give her/his opinion concerning for how long and by how much the meter has been operating outside the prescribed limits. For gas, the meter is deemed to have been registering incorrectly for the whole period since the last actual meter reading. You should argue that this should be resolved in your favour. For example, if the gas meter was over-registering, you should receive the entire extra amount charged, but if the gas meter was under-registering, you should argue that you should only pay that part which exceeds the 2 per cent limit of variation.

Older gas meters may run fast – ie, they may over-register the amount of gas consumed. This is because they use a leather diaphragm to measure the amount of gas used and this can dry out. Since 1 April 1981, gas suppliers have installed only meters with synthetic diaphragms, which are more reliable. If you have an older meter and you suspect it is recording inaccurately, you can refer it to a meter examiner. New meters may be distinguished from old ones as they have either a yellow label with a large 'S' on the meter casing or a reference number which begins or ends with an 'S'. Often the supplier will simply replace the old meter since it is recognised that leather diaphragm meters may be prone to drift into over-reading after a substantial period of years. Note that if a meter is removed by the supplier because it has made an allegation that it has been tampered with (see Chapter 9), it is important that the meter is preserved so that it can be inspected after removal. Each supplier sets out in its relevant code of practice how long it will keep a meter in such circumstances before destroying it. Check that the code of practice is being followed.

---

4. **Meter faults and faulty appliances**

Meter faults

To check whether a meter's circuit or installation is faulty, turn off all appliances (including pilot lights) and see if the meter is still registering. If an electricity meter is still registering, there may be a short circuit or a leak to earth. If a gas meter is still registering, there may be a gas leak. Apart from the effect either of these situations can have on the level of the bill, they are both dangerous and should be dealt with immediately. Once everything is turned off and the meter has ceased to register, each appliance can be checked to make sure it is registering a reasonable level of consumption by using the formulae given for checking the meter (see p82).

In rented accommodation, landlords are nearly always responsible for gas piping and electrical wiring. If you have told your landlord about defects in installations or in wiring/piping, you can ask her/him to fix the problem and to pay the difference between a high bill and the normal level of the bill, if the difference is down to the defects.

Faulty appliances

If a fairly new appliance is faulty and uses more fuel than it should, you can claim some of the excessive bill from whoever supplied the appliance by using your rights under the Sale of Goods Act 1979. Under this Act, there are a number of promises incorporated into the contract between you and the supplier of the appliance, including that the appliance is of satisfactory quality. The appliance is of **'satisfactory quality'** if a reasonable person would regard it as such, taking into account how it was described, its fitness for the purpose for which it is normally supplied, its appearance and finish, freedom from minor defects, and safety and durability.

Liability for a breach of the Sale of Goods Act lies with the seller of the goods (ie, with whom you make the contract), not with the original manufacturer. The seller is liable for faults which are present at the time of sale, whether s/he knows about them at the time or not. However, manufacturers may also be liable on occasion where a manufacturer's guarantee is included with the sale contract or under the law of negligence where faults in the goods cause damage to either individuals or to property (see p87).

The Supply of Goods (Implied Terms) Act 1973 puts similar terms into a hire purchase agreement. If an installation (eg, central heating) is installed defectively, you can use your rights under the Supply of Goods and Services Act 1982. These provide that work must be undertaken with a reasonable degree of competence and skill. If it is not, the person providing the service is liable. There are also Gas Safety (Installation and Use) Regulations covering the installation of gas fittings (such as meters and pipes) and gas appliances (such as for heating or cooking).[25] Any gas fitting installed must be soundly constructed and not made of lead or lead alloy.[26] No gas appliance can be installed unless it can be used without danger to anyone and this has been checked by the installer.[27]

If goods are bought on hire purchase or on credit for £100 or more, and the credit was supplied by a lender associated with the supplier (this includes credit cards), then the lender of the money is liable for faults as well as the supplier under the Consumer Credit Act 1974.[28] This means you can sue or threaten to sue the credit card company or other lender – this tactic can be used to put pressure on the supplier to settle any dispute or to get redress if the supplier has gone out of business.

If damage is caused by a faulty appliance, you may be able to claim compensation from the manufacturer under the general law of negligence or under the Consumer Protection Act 1987. If the damage includes physical injury to someone, seek legal advice.

5. **Overcharging by prepayment meters**

Before April 2010 suppliers could charge higher prices to customers using prepayment meters than those paying by direct debit or quarterly billing. These premiums can no longer be charged but it is possible that any household with a prepayment meter has been charged too much at some point. You have, in theory, up to six years to seek a refund. Contact Citizens Advice consumer service if you think this applies to you.

Notes

1. **Amount of the bill**
1 www.ofgem.gov.uk/ofgem-publications/85375/simplerclearerfairerfactsheet.pdf
2 Condition 21B.5 SLC
3 Condition 21B.7 SLC
4 Condition 21B.8 SLC
5 Condition 21B.4 SLC
6 Condition 21B.4 SLC
7 Condition 21B.2 SLC
8 Condition 23.4 SLC

2. **Accuracy of the bill**
9 Sch 6 para 1 EA 1989; Sch 2B GA 1986
10 *Maritime Electric Co Ltd v General Dairies Ltd* [1937] AC 610

11 British/Scottish Gas, E.ON, EDF Energy, npower and Scottish Power are committed to the Code. Scottish and Southern Energy has a Domestic Customer Charter which includes similar standards of service for billing and back billing.
12 Energy UK, www.energy-uk.org.uk/publication.html?task=file.download&id=4991
13 Limitation Act 1980
14 s6 Prescription and Limitation (Scotland) Act 1973
15 http://ico.org.uk/for_the_public/personal_information

3. **Accuracy of the meter**
16 Sch 7 EA 1989; s17 GA 1986
17 Sch 7 para 10(2) EA 1989; Sch 2B para 3(3) GA 1986

18 Sch 6 para 1 EA 1989; Sch 2B para 6 GA
    1986
19 Reg 3 EG(SP)S Regs
20 Reg 4 EG(SP)S Regs
21 Reg 8 EG(SP)S Regs
22 Reg 5 EG(SP)S Regs
23 Reg 8 EG(SP)S Regs
24 Regs 9 EG(SP)S Regs
25 GS(IU) Regs
26 GS(IU) Regs
27 GS(IU) Regs
28 s75 Consumer Credit Act 1974

Chapter 7

··

Arrears

This chapter covers:

1. What are arrears

You are in **'arrears'** of electricity or gas if you are liable to pay a bill but have not paid it on demand. You risk disconnection of your:

- **electricity** supply if you do not pay within 28 working days of receiving your bill;
- **gas** supply if you do not pay within the 28 days following the date of your bill.

The provisions discussed in this chapter only apply to arrears for charges due for the supply of electricity and gas, and not for any other purchases you have made from suppliers, for appliances or other services. Always check that you are liable for the arrears (see Chapter 5). If your bill is high, because of billing delays or incorrect reading of your meter, see Chapter 6. This chapter looks at the position where charges for the use of gas and electricity have been correctly incurred, but you have been unable to pay within 28 days.

2. Protection when you are in arrears

Standard Licence Condition 26: pensioners, the chronically sick and disabled

Under Standard Licence Condition (SLC) 26, provision is made for gas and electricity customers who are of pensionable age, chronically sick or disabled. If you fall into one of these categories, the following services may be available from suppliers:[1]

- a password may be agreed with you so that you can safely identify any person representing the supplier who visits your home;
- your bill or statement of account may be sent to any other person whom you nominate;
- help with reading a meter each quarter if you need it;
- a prepayment meter may be moved if you have reduced mobility and cannot access it.

If you are blind, partially sighted, deaf or hearing-impaired, the supplier must provide information on bills and charges that is accessible to you (eg, Braille) and provide facilities, free of charge, which enable you to ask or complain about any bill or statement of account.[2]

The Priority Services Register

Suppliers are obliged to establish and maintain a Priority Services Register listing customers who are of pensionable age, disabled or chronically sick; and who have requested to be added to the Priority Services Register.[3] Someone can make a request on your behalf for you to be added to the register. If you are on the register, you could get:

- free advice on using gas and electricity;
- a password protection scheme – anyone calling at your home on behalf of the supplier will use the password to prove who they are;
- a prepayment meter moved to a more accessible location if it is safe to do so;
- a free quarterly meter reading if you are unable to read your meter;
- bills sent to a friend, relative or carer so s/he can help to check it on your behalf;
- extra help if a gas supply is disrupted if all adults living in your property are eligible for the Priority Services Register;
- advance notice if an electricity supply has to be interrupted;
- meter readings and bills provided in a suitable format: braille, large print, audio tape, textphone or typetalk.

In addition, you may be offered a free, annual gas safety check of appliances and other gas fittings if you are eligible for the Priority Service Register, own your home, receive an income-based benefit and:

- live alone; *or*
- live with other adults, all of whom are eligible; *or*
- live with others, at least one of whom is under five years old.

Standard Licence Condition 27: difficulty in paying

SLC 27 requires that suppliers produce and publish codes of practice setting out their procedures for customers who have difficulty in paying. You should obtain an up-to-date copy of your supplier's code of practice, as they vary from one supplier to another. Suppliers must publish their codes of practice on their websites and provide copies on request.

The code of practice represents the stated policy of the supplier. It is not legally enforceable in individual cases, although a departure from the published code at policy level may be a breach of the relevant licence condition. Individual breaches should be reported to Ofgem or Citizens Advice consumer service (see Chapter 14).

The licence conditions state that suppliers must take certain steps when dealing with customers in arrears. In particular, they must provide protection for customers who 'can't pay' due to low income or inability to cope, as opposed to those who 'won't pay'. Further provisions offer protection from disconnection for vulnerable groups such as those over pensionable age, who should not be disconnected during the winter months. In practice, suppliers typically install prepayment meters instead of disconnecting supply, although this may result in consumers effectively disconnecting themselves if they lack the money to pay.

The supplier must offer a number of services when it becomes aware or has reason to believe that you are having *or will have* difficulty paying all or part of the charges for the supply of fuel.[4] The following circumstances could indicate that there is a need for such assistance:
- high consumption (over £559 for electricity or £625 for gas per year);[5]
- a sudden increase in usage;
- arrears equivalent to more than one quarter of your usual consumption;
- multiple priority debts – eg, rent, council tax or water arrears;
- payment by Fuel Direct or cash (prepayment or budget scheme);
- a history of struggling to pay or self-disconnection;
- you live in a target area as defined by the fuel poverty index or indices of social deprivation.

SLC 27 requires suppliers to take a proactive approach and act before arrears accumulate where they are anticipated. This includes:
- using Fuel Direct, where available (see Chapter 11);
- accepting payments by regular instalments calculated in accordance with an agreed plan and paid other than by a prepayment meter;
- installing a prepayment meter;
- providing energy efficiency information.

It may be useful to quote these provisions when negotiating with a supplier.

Energy UK

Energy UK is the trade association for the gas and electricity sector. Its membership includes the main energy suppliers in Great Britain – British Gas, EDF Energy, Npower, E.ON, Scottish Power and Scottish and Southern Energy. Under the

Energy UK **'safety net policy'**, no vulnerable customer should be disconnected from her/his electricity or gas supply at any time of the year. You are considered to be vulnerable if you are unable to safeguard your personal welfare or the personal welfare of other members of your household because of 'age, health, disability or severe financial insecurity'.[6]

The safety net policy compliments and supports the regulations contained within suppliers' licences. Although not legally binding, the policy could be used as a negotiating tool to avoid disconnection if you believe that you should be treated as vulnerable. Examples of what constitutes **'vulnerable'** for the purpose of the code include:

- you care for an elderly person in your household;
- you have a disability or a chronic health condition;
- the household includes young children;
- a carer, social worker, health visitor or other professional has indicated that someone in the household is vulnerable.

The safety net encourages suppliers to work with charities and advice agencies to identify vulnerability.

If you are not identified as a vulnerable customer until after the disconnection has taken place, you should be reconnected as a priority. The safety net identifies 24 hours as being a reasonable timescale, or sooner if you have a smart meter.

For more information see www.energy-uk.org.uk/files/docs/Disconnection_policy/Sept15_EUK_Safety_Net.pdf.

---

3. **Arrears in another person's name**

You may not be liable for an electricity or gas bill which is in another person's name – eg, if your partner was previously responsible for the bill and has left home, if you are a joint tenant or sharer, or if the person responsible for the bill has died. In most cases, it is possible to reach a settlement by way of a new contract between the supplier and the person taking over responsibility for supply.

You cannot be held liable for the bill while there is a genuine dispute about liability – eg, where you have been charged for supply incurred by the previous occupier of your premises.[7] See Chapter 5 for who is liable for a bill. Under rules on joint and several liability, the supplier can try to reclaim the whole amount from one or all parties to an agreement. Any discount is at the supplier's discretion.

4. **Arrears as a result of estimated bills**

Estimated bills are a common way to accrue arrears. The former Energywatch indicated that where gas arrears built up over an extended period because of a succession of estimated bills, repayment of the arrears could be made over an equivalent, extended period if you would otherwise be caused hardship. This remains a sound principle and, for example, if a meter has not been read for two years, you should be allowed two years to repay the resulting bill.

The same position could be used as a reasonable basis for making repayment arrangements with electricity suppliers. If you cannot afford the rate of repayment of arrears on this basis, you should be allowed to repay the arrears at a rate you can genuinely afford (see p95). Your supplier may request that a prepayment meter be fitted as a way to manage arrears, particularly if you are only able to offer very low repayments. If you are experiencing genuine financial hardship and would prefer not to have a prepayment meter, your supplier should also offer you a range of other options for repayment. If it does not, contact Citizens Advice consumer service for further advice. Also check your supplier's codes of practice on the payment of bills and on prepayment meters.

Where arrears have accrued because of estimated billing, ensure that your supplier has adhered to the minimum standards for meter reading (see p83). Energy UK's voluntary *Code of Practice for Accurate Bills* may also be useful. It is advisable to take your own readings regularly – at least once every three months – to ensure that a supplier's estimate of your consumption is correct. Suppliers usually amend estimated bills if you give them your own reading.

The introduction of smart meters should, in theory, see the end of estimated billing as the meter sends information to your supplier remotely about how much energy you have used and allows accurate bills to be produced on a regular basis.

Previous periods of consumption

If arrears have arisen because of a supplier's failure to read your meter accurately, you may wish to dispute your liability for the full amount of the arrears. See Chapter 6 for information on high bills.

If you need information on an earlier period of consumption concerning an estimated bill, ask your energy supplier for information on previous readings and the amount of energy used. Your supplier is required to comply with your request as soon as reasonably practicable.[8]

5. **Paying your arrears**

Methods of paying

As the supply of energy is governed by contracts, it is open to you and your supplier to negotiate and find a solution to your arrears. However, because of the large number of customers of each supplier, energy companies may find it difficult to reach individual solutions. In legal theory – and in practice – it is possible for suppliers to settle arrears by an individual payment scheme or even by writing them off, wholly or in part. This is even more likely in cases where suppliers have been at fault.

Unfortunately, it may be difficult to get a supplier to exercise any such option at first approach, since customer service staff may not be fully aware of the range of options legally available. Persistence may be necessary.

Before negotiating with your supplier, consider which way of repaying your arrears best suits your needs. Consider the practicalities of using a particular scheme and how convenient it is for you to make your payments – it is important to make an arrangement which you can keep. Chapter 4 looks at the advantages and disadvantages of the various methods of payment.

You will need to consider paying:
- through a short-term arrangement; *or*
- in instalments through a longer-term payment plan; *or*
- through a prepayment meter; *or*
- through the Fuel Direct scheme.

The ways these various options work are discussed below. The payment method available may also depend on your payment arrangements in the past. It is less likely that your supplier will agree to your choice of method if you have a succession of broken arrangements. The supplier may take the view that a prepayment meter offers the best chance of secure and regular payment without increasing arrears.

Energy companies usually expect you to reach arrangements by telephone. It is advisable to back up any conversations in writing by email and enter into correspondence wherever possible, keeping copies. Keep a record of any failure to respond to raise in your defence at a later date should the matter result in court action.

Wherever possible, deal with the supplier's complaints or customer service department, as call centre advisers often know little about the relevant law or about ways a dispute may be lawfully settled. Provide as much information about your financial situation as you can so that the staff can establish the most suitable method of payment for you.[9] Be prepared to provide information about your income and expenditure and, if you can, provide a detailed financial statement listing all your liabilities (see p96). Inform the supplier if anyone in your home is elderly, disabled or under five years old, if you claim an means-tested benefit, or if there are any other factors which cause you financial hardship, such as multiple debts.

If you are not satisfied with the options made available to you, consider using your supplier's complaints procedure. Citizens Advice consumer service can provide advice where there is a dispute about the choice of a meter or method of payment or if a deadlock situation is reached. If you are in a vulnerable situation the Extra Help Unit may help (see p180). The Energy Ombudsman may be able to intervene where the complaints system has been exhausted (see Chapter 14).

Consider applying for charitable assistance to repay or reduce arrears if you are experiencing serious hardship (see p107).

Short-term arrangements

Where problems with hardship are likely to be temporary, suppliers are often willing to come to a short-term arrangement to enable you to pay your bill in instalments, based on your ability to pay, as long as the outstanding balance is paid before your next bill arrives. If you ask for this arrangement regularly, a payment plan or a prepayment meter may be a better option.

Payment plans

Suppliers calculate an amount which you are required to pay on a weekly, fortnightly or monthly basis. This figure includes an estimated amount for current consumption and an amount for arrears.

Many suppliers add your arrears to your estimated annual consumption, and then divide by 12 for a monthly figure, or by 52 for a weekly figure. You may be told that this is the figure the computer says you have to repay in order to ensure that you will repay the arrears within a year. This is an arbitrary figure chosen by the supplier and has no legal basis. Often, the use of this formula means that you may be required to repay the arrears at a faster rate than you can afford. The overriding principle is that the method and rate of repayment should take account of your ability to pay.[10] If, for example, you can only afford to repay £3 a week, the supplier should simply add this figure to your estimated weekly consumption.

It is also important to ensure that the amount estimated for current consumption accurately reflects your use of fuel and that the supplier does not attempt to recover the arrears more quickly than you can afford by overestimating your consumption. Ensure that you read your meter each quarter so that you can accurately track your consumption and, if necessary, ask for a review of the rate of your payments.

If you have not previously had any difficulty with your bill or you have previously been able to manage a payment plan, you should not be refused this as your preferred option. If paying regularly via PayPoint is convenient for you, say so.

7

Before contacting the supplier to implement a payment plan it is useful to complete a full financial statement outlining your income and expenditure (see below). This shows you exactly how much disposable income you have. Use this information to work out how much you can afford to use to repay your energy debt. Once you have this figure, do not agree to a payment plan for more than this. This is for two reasons: firstly, you may try to meet these payments to the detriment of feeding, heating or clothing you or your family. Secondly, if you fail to meet the payments it may make your debt situation more complex, and could prejudice any future negotiations with your supplier. Therefore resist any attempts, however persuasive, by your supplier to agree to a payment plan for more than you can realistically afford.

If you do not keep to the first payment agreement made, your supplier is likely to insist you have a prepayment meter rather than renegotiate a further arrangement. However, if there are legitimate reasons why an arrangement was broken, such as a change of circumstances, this should not disqualify you, especially if it is the only or most suitable method for you.

Some suppliers may allow you to have a payment plan in conjunction with a prepayment meter set to pay for current consumption only.

Preparing a financial statement
Stage 1: Work out your income and essential expenditure

Firstly, add up all the money you have coming in from wages, benefits, maintenance and any other income every week or month, depending on how you get paid. Check that you are receiving all the benefits to which you are entitled, and that you are not paying too much tax. A free local advice agency, such as a law centre, Citizens Advice Bureau or welfare rights service can help you do this. You can also get this assistance on the telephone from National Debtline free of charge.

Secondly, work out what you spend each month on essentials. Ignore any payments for arrears at this stage. Include your normal payments for the following items:

– rent or mortgage and any other loans secured on your home;
– gas (your average weekly or monthly bill over the last year);
– electricity (your average weekly or monthly bill over the last year);
– council tax;
– water charges;
– childcare;
– transport to work;
– food;
– clothing;
– other regular household expenses – eg, telephone, internet, insurance etc.

Be realistic, and remember that payments for housing arrears are a priority (eg, rent or mortgage). Work out what you need to live on over a long period and not over a week. Do not include current payments on loans, credit agreements or catalogues.

When you have done this, deduct the total of your expenses from your total income. The difference is what you have available to deal with your debts. If this is nothing, or your expenses are more than your income, seek advice.

Stage 2: Work out your debts

Make a list of everything you owe to everyone. Include:
– arrears of rent/mortgage;
– arrears of gas/electricity/water/telephone bills;
– the total amount owing (not just the arrears) on loans, credit cards, catalogues, credit agreements, etc.
– Some debts must take priority because there are serious consequences if you cannot pay them. For most people these are:
– rent, mortgage or secured loans;
– council tax;
– magistrates' court fines;
– arrears of child maintenance.

If you are in arrears with any of these, contact the people and/or organisations you owe and try to arrange repayments. If you explain your position fully, they usually allow you a period to pay off your arrears. This may be a long period of time – particularly if you are in receipt of a means-tested benefit (see Chapter 11) or your income is very low. You may be able to reach an agreement to pay off mortgage arrears over the remaining term of the mortgage.[11] If you cannot reach an agreement, or if you think you have agreed to something you cannot afford, seek advice (see p107).

Deduct the total of what you have to pay on these priority debts from the amount you had available to pay all the debts. If there is nothing left, seek advice.

Now divide what is left fairly between all the other people you owe money to.

Stage 3: Work out how much to pay your creditors

To work out how to share this money between your creditors, there is a basic rule: the more money you owe to one creditor, the bigger share that creditor gets.

Add up all your debts (except the priority ones you dealt with at Stage 2).

Next, work out what percentage of your total debt is made up by each individual debt. For example, if your total debt is £2,400 and you owe British Gas £120, the percentage of the total debt owed to British Gas is:

$$\frac{£120}{£2,400} \times 100 = 5\%$$

Take that percentage of the weekly or monthly amount you have available to pay your debts.

In the example above, if you have £15 a month available for debts, you should pay British Gas 5 per cent of that: £15 x 5% = £0.75 a month.

Once you have worked out all these details, contact all of your creditors. They all need to see your financial statement to understand why you will only be making a small payment to each of them. Fuel suppliers ought to accept whatever you can afford to pay using this calculation. They may try to argue that they should be priority creditors, but they must accept what you can reasonably afford to pay. Your electricity supplier may have adopted a policy of accepting low rates of repayment on a *pro rata* basis with other creditors when revising its code of practice. Check your supplier's code of practice.

Prepayment meters

Prepayment meters can be calibrated to pay arrears over a period of time. For many prepayment meters, this means adjusting the meter to reclaim a fixed amount of arrears each week, irrespective of the amount of fuel used. A timing device in the meter registers the amount due towards the arrears each week. This amount is then deducted from the value of fuel paid for by inserting tokens, cards or keys, either when the meter is recharged or over the week. If you do not feed the meter every week, a build-up of these charges may result. Your supply is effectively disconnected until you can afford to pay these charges through your meter. This is known as **'self-disconnection'**.

The gas Quantum meter and similar 'smart card' electricity prepayment meters are more sophisticated in the way they are able to recover arrears. They recover an agreed fixed sum once each week and leave you with a certain minimum percentage of your credit – typically 30 per cent – for your current fuel supply. This means you always get some fuel for each credit you make.

The E.ON power card meter is set so that, where there is a build up of weekly fixed charges, you are guaranteed the use of only 10 per cent of any credit you make for your ongoing supply. The remaining 90 per cent is used towards this part of your debt. If your supplier offers this, or a similar meter, you can find out the level of this setting from the meter itself. The information booklet supplied with the meter should give details of how to obtain information from the meter. These settings can be changed by the supplier. It may be worth pressing for a change in the settings if you are facing hardship as a result of the amount you are paying back.

Smart meters can be programmed to operate in prepayment mode. You can agree with your supplier to pay the way that best suits you. In theory, they offer greater flexibility than the old style meters. For example, you can top-up credit remotely, without having to go out and buy credit, reducing the risk of 'self-disconnection' because the shops are shut.

A prepayment meter can sometimes be a good option for those with energy debts. The advantage is that you pay-as-you-go, and cannot therefore run into more debt. It also means that once the debt is paid off the situation cannot resurface in the future. However, the disadvantages of a prepayment meter include the risk that if you are in serious financial distress the lights may literally go out. See pp49–51 for more about prepayment meters.

It is rare for a request for a prepayment meter to be refused. If it is, contact Citizens Advice consumer service. If you cannot have a prepayment meter installed for safety reasons, negotiate a payment plan or request Fuel Direct if you get a qualifying benefit.

If you have previously not been able to manage a payment plan, you may be offered a prepayment meter as your only option. If this is not convenient for you, try to renegotiate another payment plan – Standard Licence Condition 27 sets out a number of alternatives including Fuel Direct.

You are not normally charged the cost of repositioning a meter to enable a prepayment meter to be fitted in these circumstances.

Resisting a prepayment meter

Suppliers sometimes attempt to impose a prepayment meter – eg, if a payment plan breaks down or if you are unwilling or unable to pay a security deposit.

– In individual cases, Ofgem has a duty to make decisions about the reasonableness of the request for security, including the request for a cash security deposit or the imposition of a prepayment meter as an alternative.

– Check the supplier's code of practice and point out its obligations. Try to negotiate an affordable payment plan in the first instance. Ensuring that a payment plan is affordable reduces the chance of it failing and avoids more difficult negotiations to reinstate a revised payment plan.

– Where a supplier asks for a security deposit, you may need to show that you are able to manage a payment plan, perhaps by referring to other bills you have successfully managed to pay in instalments – eg, catalogue debts or consumer purchases. If negotiations fail, contact Citizens Advice consumer service.

– If it is not 'safe and reasonably practicable' for you to operate a prepayment meter, the supplier is in breach of its duty to supply. Contact Citizens Advice consumer service.

Collecting arrears from a previous property on a prepayment meter

There is anecdotal evidence from some parts of the country of a supplier collecting arrears accrued at customers' previous addresses at their new properties on a prepayment meter. The Electricity (Prepayment Meter) Regulations 2006[12] and the Gas (Prepayment Meter) Regulations 2006[13] allow a prepayment meter to be used to recover a sum owed to a supplier for the supply of gas or electricity, including the provision of the gas and electricity meter, 'at any premises previously owned or occupied by the customer'.[14] Note that a supplier cannot recover sums unless it had previously entered into an agreement with you, which states in writing:[15]

• your name;

• the charges you are required to pay in addition to those recovered from a previous property;

• a guarantee that the supplier has verbally provided you with details of other payment options available to you;

- the operation of the prepayment meter as regards recovery of debt and charging for ongoing consumption; *and*
- the implications are of failing to make payments.

You can cancel the agreement by giving verbal or written notice to the supplier within seven working days of receiving the written terms of the agreement. Either party can terminate the agreement on provision of 30 days verbal or written notice.

Vulnerable customers facing disconnection, should refer to the Energy UK's Safety Net policy (see p92).

Fuel Direct

The Fuel Direct scheme allows an amount to be deducted from your benefit entitlement and paid directly to your energy supplier. A variable amount is deducted for your estimated supply, based on your previous bills. A fixed amount, currently £3.70 a week, is deducted to address your arrears. To go onto the Fuel Direct scheme you must have a fuel debt and be in receipt of income support, income-based jobseeker's allowance (JSA), income-related employment and support allowance (ESA), universal credit or pension credit. In some situations, deductions can be made from contribution-based JSA or contributory ESA. See Chapter 11 for how the scheme works.

Fuel Direct should be used as an option, 'where available', when arrears have been incurred by vulnerable people.[16] A supplier may be at fault if it overlooks this option, as many people who fall within a vulnerable category are likely to be on a qualifying benefit.

If you are eligible for Fuel Direct and are willing to have arrears deducted from your benefit, request that your supplier and Jobcentre Plus implement this. Be prepared to make a complaint if the supplier will not implement Fuel Direct or if you face procedural obstacles or delays (see Chapter 14).

Waiver, full and final settlement of arrears and estoppel

As the supply of gas and electricity is covered by contracts, general principles of the law of contract may be applied to assist you. The legal principles are complex, but there is no reason why they should not be applied to arrears. Simply giving you time to pay off arrears does not deprive the supplier of the right to take recovery action for the debt, but other principles of contract law protecting you may apply.

In some cases, an energy company may be prepared to waive its rights to recover arrears, or remit a debt, either wholly or in part. A number of doctrines of contract law, including waiver, full and final settlement and estoppel may be of assistance to a consumer with arrears in seeking a solution. These principles have yet to be tested in relation to fuel supply contracts, but for information on the general principles see textbooks on the law of contract.

Legal advice should be sought before seeking a part payment or full and final settlement based on common law principles of contract; the law relates here to England and Wales and separate advice should be sought if considering settlement in Scotland.

In many cases, a supplier may prefer to reach a final settlement on arrears on commercial grounds, rather than face the cost of recovery action through the courts for a debt. Under the small claims procedure of the county court, a supplier will not be able to recover legal costs if the sum claimed is under £10,000. This often makes taking action through the county court an uneconomic recovery option. The general attitude of suppliers is that they will go to court (including the magistrates' court – see Chapter 9) to impose a prepayment meter.

Some suppliers may be prepared to accept a one-off lump sum payment of arrears as settlement to a dispute. In contract law, parties may seek to be discharged of their obligations under an agreement by varying the terms and one party accepting a lesser payment. Many civil disputes are settled privately between parties without recourse to court action by entering into an agreement to conclude matters in a full and final settlement, rather than continue to court. The rules of court encourage parties to try to reach amicable settlements to legal claims, rather than using the court to settle disputes. There is no reason why reaching an agreement to pay a lesser sum as a full and final settlement to an energy debt should not be more widely used, as it is a common practice in other areas of commercial life.

Settlement by paying a lesser sum may be an option where you are in a position to switch supplier (although the successor company may be entitled to recover certain arrears if the debt is assigned, if a full and final settlement is not reached).

However, the legal rules can be quite complicated. Simply sending a small amount to an energy supplier is not sufficient to discharge arrears or a sum owed in interest.[17] The supplier is still lawfully entitled to recover the money, unless some benefit is derived from the terms you may offer to the energy supplier, along with the part-payment.

For part-payment to represent a settlement, the supplier must formally agree to waive its right to recovery action in return for accepting a lump sum or by agreeing to drop some obligation upon itself, such as the requirement to read a meter or investigate a complaint. Alternatively, you might agree to accept a prepayment meter, varying the terms of an existing contract, in return for the supplier accepting a lesser sum in full and final settlement of the existing arrears. Where a company is threatening legal action, it may be possible to pay part of the money owed in return for the company waiving its right to take legal action (thus saving itself the cost of litigation). You must make a clear offer to the supplier that you are willing to pay a smaller sum immediately towards clearing arrears and it is necessary for the supplier to agree and to waive the right of court action. Alternatively, the energy company may agree to waive its right to recover a sum in arrears if you have switched supplier. In some cases, you might agree to waive the right to pursue a claim against the supplier or to Ofgem for a breach of the licence conditions. In such a case, the supplier will have derived a benefit (ie, not being subject to a complaint or legal claim) and may agree to accept a lesser sum. The conferring of benefits on both the consumer and company establishes a settlement to the debt.

Make your offer in writing and clearly refer to acceptance being a full and final settlement of arrears and a waiver of further recovery action.

Any communications should be marked 'full and final settlement', including endorsing this on any final payment cheque.

Example

Mrs Green was a widowed pensioner who had been a customer of one supplier for many years. Before her husband died, he switched to a new supplier. The new company failed to read the meter and sent an estimated bill of over £1,000, which was carried over each quarter.

Efforts to negotiate proved fruitless, and letters went ignored. Mrs Green contacted her previous supplier who had kept records of her good payment history. Mrs Green paid sums she could afford, and repeatedly wrote to her new supplier offering to pay by instalments. No written response was forthcoming. The new supplier then threatened legal action to recover the arrears. On taking advice, she was recommended to make a payment of 10 per cent of the bill in full and final settlement. She sent a cheque for 10 per cent of the outstanding amount and the letter was clearly marked 'cash this cheque only if you accept this sum in full and final settlement of the dispute'. She also endorsed the cheque 'full and final settlement cheque'.

The new supplier cashed the cheque after Mrs Green switched supplier, returning to her old energy company with whom she had a good payment record.

Four months after she returned to her old supplier, the new energy company presented her with a fresh demand. The letter in full and final settlement was cited in negotiations. It was argued that by cashing the cheque in response to the settlement letter the supplier had accepted the offer of settlement by Mrs Green. Eventually the supplier agreed to waive recovery and wrote off the arrears when it emerged that it had not read Mrs Green's meter for over a year.

Disputes over whether a supplier has agreed to full and final settlement may come down to a dispute of fact where the energy company has accepted payment. Legal advice should be taken in such cases.

Front line staff may not be familiar with these principles, so you should seek to speak to the complaints handling team or a senior manager.

In some cases, a supplier may agree to forego the right to recover arrears for a period and not enforce a debt in return for a lesser sum in payment. During that period, the supplier cannot go back on the agreement and take action. What is known as the doctrine of **'estoppel'** operates. This provides that where a person acts in reliance of a promise made on another, that other person cannot then break the promise and take enforcement action. For example, where a supplier agrees to grant a three-month suspension on paying arrears and you act accordingly, the supplier cannot demand the money in the three months. The supplier is said to be 'estopped' from taking action in the period.[18]

6. **Rate of repayment**

Fuel suppliers are required to make arrangements for the recovery of debts which take into account your ability to pay. This applies whether you are offered a payment plan or a prepayment meter.

The rate of Fuel Direct deductions (see p159) is often used as a yardstick to determine the period over which a debt should be repaid within a payment plan or through a prepayment meter.

If you get income support, income-based jobseeker's allowance, pension credit, income-related employment and support allowance or, in some circumstances, universal credit, you may wish to consider the Fuel Direct scheme, but note that this would result in fixed deductions for your arrears. This may not be appropriate if your arrears should be recovered at a lower rate.

Repayments below Fuel Direct rates

If your income is low, or roughly equivalent to means-tested benefit levels (eg, you receive housing benefit or council tax reduction), argue that you should not repay your debt at a higher rate than the Fuel Direct rates.

There are many situations where a supplier should accept less than this level of repayment, particularly if your income is very low – eg, you may not be entitled to benefit because you work, but you have to pay childcare or mortgage expenses which are not taken into consideration; or your benefit is sanctioned; or you have multiple other priority debts.

Suppliers are primarily concerned with ensuring that you pay for your current consumption as the first priority, and should accept payments of arrears over extended periods of time at rates which you can afford to sustain, even if this is very low. If you have multiple debts, your suppliers should be persuaded to treat fuel debts in the same way as other priority debts. If you cannot afford the rate of repayment sought, ask to pay at a lower rate. Provide your supplier with information about your income, necessary outgoings and other debts by sending a detailed financial statement (see p96). If the supplier refuses to accept lower payments, contact Citizens Advice consumer service for advice.

If a supplier formally refuses a particular repayment rate but then goes on subsequently to accept regular repayments from you at that rate, it is arguable that an agreement by conduct has been reached. Similarly, if a supplier has promised not to take enforcement action or pursue a certain sum and you then act in reliance on that promise, it is arguable that the supplier cannot then renege on the promise. For example, if a supplier tells you that it is prepared to reduce part of a debt and you then repay the remainder, it should not be open to the supplier to take action regarding the unpaid balance at a later date. The rule on estoppel (see p102) applies.

Arrears more than six years old

Where arrears for fuel accrued more than six years ago, recovery is statute barred under the Limitation Act 1980, which imposes a six-year time limit on the recovery of contractual debts in law. The six years run from the date that the bill first fell due. Suppliers cannot instigate legal action after this date, although this does not stop them from serving a demand for a bill which is more than six years old. However, you are not under a legal obligation to pay if faced with a demand which relates to a debt more than six years old. Where legal action has been commenced within the six-year period (eg, the supplier has taken court action against you and obtained a judgment), then they can enforce the judgment even after six years but may need to seek the permission of the court to do so, and the court would take into account any failure on their part to act promptly. If the debt accrued under back-billing rules (see p78), seek legal advice.

Paying for your current consumption

Whatever your rate of debt repayment, you will also have to pay for your estimated current consumption over the year. It is important that this estimate is as accurate as possible, otherwise you may end up paying more than you can afford. The requirement at Standard Licence Condition (SLC) 31A for suppliers to provide detailed information about your last year's consumption will be useful here.

You can check the estimate provided by the supplier, either by using your own bills or by asking for details of the actual readings of your meter over a past period. Suppliers often have records for up to eight quarters. Try to make sure that the readings cover at least a year so that you make allowances for seasonal variations, and any changes in your lifestyle or appliance usage. Calculate the number of units of fuel you have used over the period covered by meter readings, and then divide this figure by the number of weeks in that period to work out the number of units you use on average each week. Multiply that figure by the cost of the units of fuel. Then add on the amount of the standing charge for each week.

If you have not been in your property for long, the supplier's energy efficiency advisers should be able to advise you on the likely size of your bills if you provide them with details of the size of your home, family and the appliances you use.

Where you have a payment plan, ensure that your meter is read every quarter, either by you or the supplier, so that you can check the accuracy of your estimated current consumption and ensure that the amount you are paying is neither too low nor too high.

Lump-sum repayments

You should not be asked to pay lump sums of money as part-payment towards arrears as a condition of being allowed a prepayment meter; nor should you be asked to pay lump sums before being allowed to pay arrears in instalments if you are unable to pay your bill because of hardship. The supplier's licence specifically provides that it must allow you to pay your arrears in instalments, taking into account your ability to pay.

However, a supplier might demand a part-payment towards a debt before agreeing to install a prepayment meter. This is contrary to the licence provisions, specifically SLC 27. Refer a dispute arising from such a request to Citizens Advice consumer service.

Breakdown of repayment arrangement

If a change of circumstances has occurred which has affected your ability to pay, ask the supplier to consider a revised payment arrangement. Suppliers are obliged to consider this under the licence conditions. Your supplier may include a statement about changes of circumstances in its code of practice. Note that the provisions within SLC can be enforced by Ofgem in individual cases, in contrast to the provisions within the code of practice, which are voluntary. Any special circumstances ought to be brought to the attention of the supplier.

If your circumstances have not changed, it may have been that the level of repayment was too high in the first place. If this is so, ask for a revised payment arrangement based on a more detailed picture of your financial circumstances. For example, if you have obtained debt advice for the first time and your adviser has helped you to prepare a financial statement. Otherwise, you may be obliged to accept a prepayment meter as an alternative to disconnection.

If you cannot manage the payment level you have agreed, always try to pay something, however small, and on your usual payment date, even if you are in the process of negotiating a new payment arrangement. Doing so demonstrates 'good faith'; you want to address your debt, but cannot afford to do so at the previously agreed level.

In the event of a dispute and a threat of disconnection, contact Citizens Advice consumer service for advice.

---

7. **Switching supplier when you have arrears**

Where there is an unpaid bill, a supplier may block a transfer to another energy provider.

However, if you have a prepayment meter and there are outstanding charges, these may be assigned to a new supplier. This agreement has been formalised by a change to the Standard Licence Conditions.[19] Additionally, the 'big six' fuel suppliers have increased the debt threshold voluntarily from £200 to £500. Within this boundary your existing supplier cannot block a transfer.

Similarly, if you have a disputed debt of £500 or less transferred onto a prepayment meter, the supplier cannot block a transfer where charges are disputed in their entirety or there is supplier error. The precise scope of this condition has yet to be tested by the courts or by the regulators.

One such situation may be where you have offered and paid a lesser sum in full and final settlement for an existing debt and accepted the imposition of a prepayment meter.

Suppliers are required to:

- keep evidence of that request and of the reasons for it for at least 12 months after the request is made; *and*
- as soon as reasonably practicable inform the proposed new supplier:
  - that the objection has been raised at your request; *and*
  - of the reason given by you for making the request.

Normally, within 90 days of taking over a supply, the new supplier is given a notice by the previous supplier to assign the charges. The new supplier must then pay the outstanding charges to your previous supplier. Your new supplier can then seek to recover the money from you. If the old supplier fails to act within prescribed time limits, the right of assignment may be lost.

In such cases, you should not be regarded as having failed to pay any charges for the supply of fuel:[20]

- if the supply is genuinely in dispute; *or*
- where charges are in relation to the provision of a gas meter.

Contact Citizens Advice consumer service for advice on the position with charges and difficulties with a supplier; it may also be worth involving the Energy Ombudsman (see Chapter 14).

Ultimately, it is the county court in England and Wales and the sheriff court in Scotland which could be called upon to determine the matter (see Chapter 14). A debt which has been assigned is not enforceable if there is a dispute which you are prepared to raise as a defence. Only when the dispute has been resolved against you and a judgment given will any debt be legally enforceable. Alternatively, your new supplier also has discretion to waive charges or collect a lesser sum.

8. **Multiple debts**

For many people, fuel arrears are just part of a bigger problem of unpaid bills. It is beyond the scope of this book to give detailed advice about debts other than those for fuel. See CPAG's *Debt Advice Handbook*.

If you are seeking advice about your fuel debts, tell the adviser about all your other debts as well, so that s/he can provide you with accurate and relevant advice.

Getting advice

Your local Citizens Advice Bureau can tell you what is available locally and can advise on multiple debts. There is also a national telephone helpline, National Debtline (see www.nationaldebtline.org). The Money Advice Trust produces a comprehensive self-help guide for dealing with multiple debts. The guide includes a very useful sheet to help you work out your own financial statement. Copies of this pack can be obtained from your local free advice agency.

You can also consult solicitors, law centres or advice agencies which provide debt advice. Certain charities also operate assistance schemes. Make sure that any advice you get is free and genuinely independent of your creditors.

Charitable assistance

Many of the fuel companies offer charitable assistance in the form of energy trusts to customers experiencing financial difficulty (see p181). Check whether your fuel supplier runs such a scheme.

The aim of these schemes is to free individuals from fuel debts and enable them to make a fresh start. You can only ask for help to clear arrears – you are usually expected to set up a regular payment arrangement to manage your future supply and prevent arrears accruing on your account again. The qualifying criteria for these trusts are relatively broad. You must be able to show that you are experiencing financial hardship, and that a payment from the trust will assist you to pay your future fuel bills. Not all applications are successful and it can take a number of weeks for the trustees to decide about your case. The decision made by the trust is final and there is no right of appeal. Make sure that you give as much information as possible about your circumstances including why you have had difficulty paying your energy bills. You will also be required to complete a detailed financial statement listing your other commitments and any other debts you may have. Some trusts offer further help in the form of 'further assistance payments' to meet the cost of either applying for bankruptcy or a debt relief order. These can be useful if you have multiple debts and want to pursue either of these options but cannot meet the cost of doing so yourself. Check whether your supplier offers this type of help.

Administration orders

If you have multiple debts, you can apply to a county court in England and Wales for an administration order as a way of getting the county court to take over the administration of your debts.[21] This is a little used, but extremely useful, provision whereby the court will decide how much you can afford to pay to each of your creditors. You make one monthly payment to the court, which then pays your creditors on your behalf.

To be able to obtain an administration order:

- you must have a county court or High Court judgment against you;
- the total of your debts must be less than £5,000;
- you must have at least two debts.

Administration orders are not available in Scotland.

Importantly, the court can reduce the amount of overall debt paid by order so that only a percentage of the total debts are to be paid – eg, ordering that the debtor pays into the administration order 25p for each £1 owed. Thus, a person with total debts of £4,000 would only be required to ultimately pay back £1,000 under the administration order.

In terms of fuel debt, the making of an administration order by a court must be the definitive statement of your ability to pay. Suppliers who seek to recover arrears outside the terms of the administration order could find themselves in contempt of court.

Suppliers, however, are not used to people putting their debts into administration orders. Always contact your supplier in advance to explain your proposed course of action. You will also need to make arrangements to pay for your current supply. The supplier will almost certainly insist that you pay using a method which offers the supplier greater security, such as a prepayment meter, Fuel Direct or a payment plan using direct debit. The supplier is not permitted to recover arrears through any of these methods of payment without the leave of the court, since the court will take over the payment of arrears.

Apply for an administration order on form N92 available from your local county court or www.gov.uk. The application requires details of your means, whether you are employed or unemployed and details of all your creditors and how much you owe them. When you have completed the form, take it back to the court, where you have to swear that the contents of the form are correct. The court then contacts your creditors and either makes the order by agreement or arranges a private hearing with a district judge to consider your application. In practice, few creditors bother to attend a hearing for an administration order, and in many cases they may write off the debt completely at this stage.

The court has powers to cancel or vary the administration order once it is granted, if you fail to pay. However, if you do encounter payment problems, contact the court immediately to seek a variation of the order.

Advantages of an administration order

- The order usually runs for a period of three years, though this is not automatic (check with the court). Provided you have paid your monthly payments, the rest of the debt is written off at the end of this period.
- Interest is frozen on accounts.
- Your creditors cannot chase you or take other court action against you while the order runs.
- You have to make only one payment each month.
- The court can reduce the amount of each debt owed by requiring only a percentage of the debt to be paid.
- It is possible for a charity or a third party to pay off the amount owing under the administration order in a lump sum.

Disadvantages of an administration order

- Your name appears on a register of court orders and you may find it hard to get credit.
- There is a court fee each time you make a payment. This cannot be more than 10 per cent of your debt – eg, if you owe £4,000, the total fee must be a maximum of £400.

Debt relief orders

In England and Wales, if you are unable to pay your debts and you meet the eligibility conditions, you can apply for a debt relief order (DRO).

To qualify, you must satisfy the following.[22]

- You must be unable to pay your debts.
- Your total debts must be no more than £20,000.
- Your total assets must not be more than £1,000 although you are also permitted to additionally own a vehicle worth up to £1,000 (or more if it has been adapted because you have a disability).
- Your disposable income after deducting allowable expenses must not be more than £50 a month.
- You do not own your home.
- You have lived, worked or had a property in England or Wales at any time during the last three years.
- You must not have been subject to a DRO within the last six years.
- You must not be involved in any other formal insolvency procedure at the time of application for a DRO.

Under a DRO, recovery action by creditors to obtain money from you is stopped and at the end of a set period (usually a year) the debts are cleared. So, for example, a supplier cannot take any further action against you such as disconnection in respect of fuel arrears included in a DRO.

Certain debts are excluded from a DRO. See CPAG's *Debt Advice Handbook* for further details. Although you can include fuel arrears in the order, you are expected to meet the costs of future fuel consumption and should include these in your expenditure figures.

To apply for a DRO a fee of £90 is payable. This can be spread over six monthly instalments.

DROs are administered by the Official Receiver through the Insolvency Service. You can only apply online but only through an approved third party or intermediary. These are usually debt advisers who have authority to complete the application forms. To find your nearest approved intermediary, contact your local free advice agency.

Visits from enforcement agents

In England and Wales, if you have fuel debts enforcement agents – formerly bailiffs – (in Scotland, sheriff's officers) may come to your home to take away your possessions. Enforcement agents cannot force entry to private dwellings in England and Wales for most types of debts, and in Scotland forced entry is only possible as a last resort. If a company or its enforcement agents threatens to force entry, make a formal complaint to the supplier. Under rules introduced by the Enforcement Act 2007, in England and Wales, from 6 April 2014 enforcement agents must give you seven days' notice that they will be calling at your property.

Forced entry to your home in connection with an unpaid energy bill can only be by warrant of control (see Chapter 10). Even if enforcement agents gain entry and seize goods, the sums of money raised at auction rarely cover the bailiff and auctioneer fees and the debt remains.

Enforcement agents have a duty to notify the creditor and report the circumstances in situations where there is evidence of vulnerability. The Ministry of Justice has produced standards as to what constitutes 'vulnerable situations' (see Appendix 4).[23]

In Scotland, sheriffs cannot demand entry to your home unless a court order known as an 'exceptional attachment order' has been obtained. Forced entry cannot take place unless there is a person present who is at least 16, and is not, because of her/his age, knowledge of English, mental illness, mental or physical disability or otherwise, unable to understand the consequences of the procedure being carried out.[24]

Energy suppliers do not have powers to force entry simply to recover money – any power of entry can only be exercised under a warrant through a magistrates' court to disconnect a supply, not to seize possessions (see Chapter 10).

Notes

2. Protection when you are in arrears
1 Condition 26.1 SLC
2 Condition 26.2 SLC
3 Condition 26.4 SLC
4 Condition 27.5 SLC
5 www.gov.uk/government/uploads/
system/uploads/attachment_data/file/
524704/Domestic_energy_bills_in_
2015_-_The_impact_of_variable_
consumption.pdf
6 Energy UK, *The Energy UK Safety Net: Protecting Vulnerable Customers from Disconnection,* February 2016

3. Arrears in another person's name
7 Sch 6 para 2 EA 1989

4. Arrears as a result of estimated bills
8 Condition 21B.9 SLC

5. Paying your arrears
9 Condition 27 SLC
10 Condition 27.8 SLC and codes of practice from suppliers
11 *Cheltenham and Gloucester Building Society v Norgan* [1996] 1 All ER 449
12 SI 2006 No.2010
13 SI 2006 No.2011
14 Reg 3(1)(a) E(PM) Regs; reg 3(1)(a) G(PM) Regs
15 Reg 4 E(PM) Regs; reg 4 G(PM) Regs
16 Condition 27.6 SLC
17 *Foakes v Beer* [1881-85] All ER 106; *Pinnel's Case* (1602) 5 Co Rep 117
18 *Central London Property Trust v High Trees House Ltd* [1947] 1 KB 130

7. Switching supplier when you have arrears
19 Condition 14.6 SLC; Ofgem debt assignment protocol for prepayment meter customers, letter 24 September 2012
20 Sch 6 para 1 (9) EA 1989; Sch 2B para 2B GA 1986

8. Multiple debts
21 s112 County Courts Act 1984; CCR 39
22 Sch 4ZA Insolvency Act 1986
23 Ministry of Justice, *Taking Control of Goods: National Standards,* 6 April 2014

24 s18 Debtors (Scotland) Act 1987; s49 Debt Arrangement and Attachment Act (Scotland) 2002

Chapter 8

Disconnection for arrears

This chapter covers:
1. When you can be disconnected for arrears (below)
2. Protection from disconnection (p117)
3. Preventing disconnection (p118)
4. At the point of disconnection (p120)
5. Getting your supply reconnected (p121)
6. Disputes: unlawful disconnection (p123)

1. When you can be disconnected for arrears

The most important power which suppliers have for non-payment of energy bills is to cut off your supply if you do not pay your bill. The power to disconnect is considered more effective than the right to recover money through the court system.

Unfortunately, suppliers sometimes find it difficult to distinguish between deliberate non-payment and those who would pay but are suffering financial hardship or other problems. However, disconnection should only be a last resort in extreme circumstances, when all other methods of recovery have failed. It should not be used as a standard method of debt enforcement.

In practice, very few customers are now disconnected as a result of a failure to pay bills. According to Ofgem, at 31 December 2014 there were an estimated 1.4 million electricity accounts in arrears with an average debt of £355 and an estimated 1.2. million gas accounts in arrears with an average debt of £382.[1]

The declining number of disconnections

In 2003, there were 15,973 disconnections of gas and 1,361 for electricity.[2]
Between January and December 2014, 192 electricity disconnections and 42 gas disconnections took place in the UK.[3] This represents a very small percentage of the number of customers in fuel debt during the same period – 1,362,961 electricity customers and 833,135 gas customers – and an even smaller percentage of all customers nationally.

Informing the supplier of vulnerability

Suppliers must conform to Standard Licence Condition (SLC) 27 with regard to vulnerable customers who fall into arrears (see p117). This condition covers payments, security deposits, disconnections and final bills.

It is important as a preliminary step in all cases to discover the cause of your arrears, and identify whether you fall into a vulnerable group. If someone in your household is vulnerable, inform your supplier as soon as possible, preferably before arrears arise.

Energy UK's definition of vulnerability

A customer is vulnerable if, for reason of age, health, disability or severe financial insecurity, s/he is unable to safeguard her/his personal welfare or the personal welfare of the household.[4]

If you fit within this definition you are likely to be in receipt of one of the following benefits:
- retirement pension;
- pension credit;
- disability living allowance;
- personal independence payment;
- attendance allowance;
- long-term incapacity benefit;
- employment and support allowance;
- income support (IS);
- IS with disability premium;
- income-based jobseeker's allowance;
- universal credit.

If you are entitled to such benefits but there has been a problem with your claim or your benefit has been sanctioned or suspended, quote these provisions in initial correspondence with a supplier.

Potentially vulnerable categories include those listed in paragraph 77 of *Taking Control of Goods: National Standards* (see Appendix 4) as well as those included under SLC 27 (see p117). Suppliers may also have their own definitions.

Remember that individual circumstances can vary, and that people may move in and out of vulnerable categories, or be vulnerable under more than one heading.

Your supplier is required, at least once a year, to take 'all reasonable steps' to inform all customers about the priority services register (see p90) and how you can be listed on it if you are of pensionable age, disabled, have a hearing or visual impairment or long term ill-health.[5] If you have different suppliers for gas and electricity, you need to register separately with both.

If the supplier does not respond properly to the information about vulnerability, ignores information it holds, or if there is delay, you should consider making a formal complaint (see p116). Also inform Citizens Advice consumer service which may refer you to the Extra Help Unit (see p180).

When the supplier can disconnect – electricity

Contract suppliers can only disconnect for arrears if the contract says so. If you are threatened with disconnection, check your contract carefully.

With any remaining tariff customers, a tariff supplier may disconnect your supply if you have not paid all charges due in respect of your electricity supply[6] within 28 working days after the date of the bill or other written request to pay.[7]

You must be given at least seven working days' written notice of the intention to disconnect[8] and cannot be disconnected for any amount which is 'genuinely in dispute'.[9] '**Charges due**' include any amounts for the electricity supply to any premises, including standing charges, meter provision, the provision of an electrical line or plant or sums due under a green deal plan.[10] They do not include other charges such as those for credit sale agreements or for repositioning/ adapting a meter for a disabled person.[11]

Your supplier might attempt to wrongly include charges for periods when supply was disrupted and was not reconnected in a reasonable time. Check carefully that you are only being asked for the arrears for which you are liable.

Charges due can only be properly established on the basis of a meter reading. Estimates cannot be used, and you should not be disconnected on the basis of an estimated bill. However, you must ensure you pay the amount you agree you do owe; otherwise there is an increased risk that you may be disconnected. The supply can be cut off at the premises to which the bill relates.[12] Failure to give the required notice of disconnection is an enforcement matter. Contact Citizens Advice consumer service for advice urgently if you think your supplier intends to disconnect your supply.

When the supplier can disconnect – gas

Gas suppliers supplying under the terms of contracts and deemed contracts may disconnect your supply if you have not paid any charges due for gas within the 28 days following the date of the bill.[13] '**Charges due**' are any charges in respect of the supply of gas.[14] You are entitled to seven days' notice in writing of the intention to disconnect.[15] This is usually given in the final demand, which may arrive earlier than the 28 days above. A gas supplier is not entitled to disconnect your supply for any amount which is 'genuinely in dispute'.[16]

It is important to pay any undisputed part of the bill, as well as to maintain or establish a payment arrangement for ongoing fuel costs while reaching a resolution of the dispute.

Note that public gas transporters may also disconnect your supply in certain circumstances.

Disconnection for assigned arrears

If you have changed supplier and you owe money to your previous supplier, the previous supplier can assign some of its debt to your new supplier in certain circumstances (see p105). The new supplier may cut off your supply as though it were the previous supplier if you fail to pay. You are entitled to a minimum of seven days' notice of the new supplier's intention to disconnect.[17] There is no right to disconnect when the entire bill is genuinely in dispute.[18] You may be protected from disconnection by conditions contained in your supplier's licence if you fall into one of the protected categories (see p117).

Disconnection when you pay in instalments

If you have an arrangement to pay in instalments, either for your current supply only or for your current supply plus an amount for arrears, the supplier is not entitled to disconnect for arrears while you keep to the terms of your agreement, since you are paying the amounts requested in writing.[19]

If you miss a payment, the supplier is entitled to disconnect for arrears from 28 working days after the date of your missed payment, but only if there are still charges due, and providing you have been given seven working days' notice of the intention to disconnect. If your account is in credit, based on a reading of your meter and allowing for any standing and other charges, the supplier is not entitled to disconnect.

In practice, rather than actually disconnect, suppliers are more likely to try to impose a prepayment meter. The supplier has to inform you of its intention to disconnect, subject to the notice period above, before being entitled to install a prepayment meter no less than 28 days from the date of your missed payment. A prepayment meter may be installed with your agreement at any time. See p99 for how to resist a prepayment meter.

Alternatively, your supplier may decide that a security deposit is required just in case you do not keep to your agreement to pay by instalments. The supplier has to write to you to inform you of this. The supplier must give you notice that it intends to disconnect if you do not pay a security deposit. A prepayment meter may be imposed if you do not wish to pay a security deposit. See p36 for more about security deposits.

If the breakdown of your instalment arrangement has occurred because you cannot afford to pay, you may be able to arrange an alternative, more suitable payment plan. In any event, the supplier must offer you a prepayment meter as an alternative to disconnection, providing this is safe and practical.

Payment arrangements fail for a variety of reasons. It might have been the wrong arrangement from the outset: the commencement date may not coincide with the receipt of income; the frequency of payments may not coincide with the receipt of income; or the amount for consumption or arrears may be set too high (or too low). There may have been a change in lifestyle or appliance use. You may have been ill or in hospital. Failing to keep to an agreement should not automatically preclude the possibility of another one being arranged – it is essential to establish why a previous arrangement failed. Suppliers are required to make instalment arrangements based on what you are able to pay.

Fuel Direct (see p159) should also be considered as a payment option. Suppliers are expected to use '*where available*, a means by which payments may be deducted at source from a social security benefit received by that customer.'[20] The use of the words 'where available' suggests that a supplier may be at fault if it overlooks or ignores this option.[21]

Complaints about disconnection

To complain about disconnection by an energy supplier, first contact the company concerned. Energy suppliers are subject to regulations setting out how to respond to a complaint (see Chapter 14).

Ofgem usually refuses to investigate a complaint until you have given the supplier a reasonable opportunity to deal with it. If you need help to make your complaint, seek advice from Citizens Advice consumer service. If your complaint has not been resolved to your satisfaction within eight weeks you can ask Ofgem to investigate your complaint. Provided you have exhausted the supplier's internal complaints procedure, Ofgem can investigate your complaint about disconnection against:

- a gas transporter for:
  - disconnection of, or a threat to disconnect, your gas;
  - refusal to reconnect your supply following disconnection;
- a gas supplier for:
  - cutting off, or a threat to cut off, your gas;
  - refusal to reconnect your supply following disconnection;
  - the failure of a prepayment system;
- an electricity supplier, distributor or licence holder for:
  - disconnection of, or a threat to disconnect, your electricity;
  - refusal to reconnect your supply following disconnection;
  - the failure of a prepayment system.

For more information on complaints, see Chapter 14.

2. Protection from disconnection

The following provisions may prevent disconnection in certain circumstances.

- If you are having difficulty paying your bill, condition 27 of the Standard Licence Conditions (SLCs – see below), the supplier's code of practice and Energy UK's safety net may offer some protection (see p92).
- If you claim certain means-tested benefits, Fuel Direct may be an option (see p159).
- The supplier may not be entitled to disconnect (see p123).
- If you are a tenant, your local authority may be able to help (see p200).

Condition 27 of the Standard Licence Conditions

If you are threatened with disconnection because you cannot pay your bill, SLC 27 gives you the following rights.

- You are entitled to a payment arrangement to repay your arrears at a rate you can afford.
- Using, where available, a means by which payments may be deducted at source from a social security benefit such as Fuel Direct (see p159).
- You may pay by regular instalments and through a means other than a prepayment meter.
- If you have not been able to manage a payment arrangement, you must be offered a prepayment meter (if safe and practical) as an alternative to disconnection. The meter must be set to recover arrears at a realistic rate which you can afford.
- If you are a pensioner or have children under 18, you should receive protection from disconnection in winter (see below). If the supplier knows or has reason to believe you are such a customer, it *must not* disconnect you if you live alone or live with another pensioner or children under 18.[22] This provision applies even where there is suspicion of theft of electricity.[23]
- If you are below pension age and your household includes persons who are of pensionable age, disabled or chronically sick, the supplier must take 'all reasonable steps' to avoid disconnecting your supply in winter.[24]
- You should be offered information about how you can reduce your charges by using fuel more efficiently.

Winter
'Winter' is defined as the months of October, November, December, January, February and March.[25]

Suppliers are obliged to develop methods for dealing with customers in arrears under the terms of this condition. These set out the procedures which should be followed by each supplier and provide the practical mechanism for protecting your rights. Any departure from the methods may constitute a breach of your rights and could be referred to Ofgem for investigation (see p1).

Calculation of an instalment rate

Under SLC 27 a supplier must not disconnect you unless it has first taken all reasonable steps to arrange the repayment of outstanding charges.

This means that your supplier 'must take all reasonable steps' to discover your ability to pay and must take this into account when calculating instalments, whether this is by way of a payment plan or through a prepayment meter.

You should provide your supplier with a detailed financial statement (see p96) setting out details of your income, expenditure and debts.

The supplier must also consider:
• relevant information provided by third parties, where it is available; *and*
• where instalments will be paid using a prepayment meter, the value of all of the charges that are to be recovered through that meter.

> *Other situations where disconnection should not occur*
> Examples of when supply will not be disconnected include:
> – if you agree and keep to a payment plan;
> – if the debt is in the name of a past customer, such as an ex-partner, and you have made arrangements to take over the supply;
> – if you have a query about your bill and you have paid the part of it which you agree you owe;
> – you have agreed to have a prepayment meter installed;
> – you request Fuel Direct;
> – you have contacted social services for help (see p178) – you must tell the supplier that you are doing this.
> It may be useful to suggest these alternatives to disconnection as basis for negotiation with suppliers.

3. **Preventing disconnection**

There are reasons why you should always try to prevent disconnection.
• You cannot solve a debt problem by being disconnected. Even after you have been disconnected, the supplier still requires you to pay the money you owe and can take court action to get it. You will also be charged a fee for the expenses of disconnecting the supply.
• If you later want your supply reconnected, you will have to pay any arrears still owing, plus the costs of disconnection and reconnection.

- If the supplier has had to get a magistrate's warrant (or Justice of the Peace or sheriff's warrant in Scotland) to disconnect your supply, you also have to pay the costs of obtaining the warrant (although an application may be made to the court to use its discretionary power to refuse the claimed costs). If you have produced a means statement (see p96), this should be submitted to the court.

You may have to pay a security deposit as a condition of being supplied in future following disconnection. If you are finding it difficult to meet your fuel costs, seek advice and let your supplier know about the difficulties you are experiencing as soon as you can. You are more likely to be able to obtain a solution which genuinely meets your needs if you have time to think about your proposals to the supplier, or what the supplier is prepared to offer you.

You should be able to prevent disconnection if:
- you contact the supplier; *and*
- you arrange to pay your arrears at a rate you can afford; *or*
- you request to be put on the Fuel Direct scheme (see p159); a request may also be made by the supplier; *or*
- you agree to accept a prepayment meter set to collect the arrears at a rate you can afford.

In practice, disconnection most often occurs where there has been no contact between the customer and the supplier. Once you contact the supplier, the supplier must consider your situation and work with you to identify a suitable way for your supply to continue and for you to repay your arrears at a rate you can afford (see Chapter 7).

The codes of practice may make some provision for disconnection to be delayed, typically for 14 or 21 days, if you tell the supplier that you are going to ask social services (social work in Scotland) for help with the bill.

Usually suppliers prefer you to contact them by telephone. It is safer in terms of establishing an agreement and ensuring that the correct information is received, to confirm everything in writing. If possible, send an email to your supplier to confirm what you have agreed to pay so that you have a record for future reference.

If all else fails you can still make representations at the warrant stage if the matter goes to the magistrates' court (see Chapter 10). It is often possible to negotiate a settlement even at this point. You should always attend court, taking an adviser or friend with you if possible.

4. **At the point of disconnection**

The supplier's right to enter your premises

The supplier may, with your consent, enter your premises to disconnect your supply, providing it has served you with a correct notice of disconnection and has published details of Standard Licence Condition (SLC) 27 on its website (see p117 and Chapter 10).[26] If you do not consent, the supplier must obtain a warrant from the magistrates' court (or sheriff court in Scotland). The costs of the warrant are added to your bill – these are generally around £75–110 and the supplier may also charge for expenses legitimately incurred.[27] Although in some cases, the supplier will disconnect your supply from the mains outside your property – which makes reconnection extremely expensive. It has yet to be tested in law to what extent, if any, a realistic offer to clear arrears may be grounds for a magistrates' court to refuse to issue a warrant of entry to disconnect, but magistrates would be expected to act reasonably and consider all the relevant circumstances.[28]

Your home must be left no less secure than it was before entry. Any damage caused by legally gaining entry must be made good or compensation paid. If the supplier fails to secure your premises and your possessions are stolen as a result, you can sue the supplier. Suppliers sometimes change locks and leave a note telling you where to pick up a key.

Disconnecting external meters

A warrant is not required for the disconnection of an external meter. Disconnection is lawful, providing the correct notices have been given (see p114).

Disconnecting smart meters

The new technology contained within smart meters means that suppliers can disconnect your supply remotely without visiting your home. Ofgem has sought to address this development by modifying and strengthening the existing protection contained in SLC 27 for all vulnerable customers.[29] Suppliers must be able to show that they have:
- taken proactive steps to establish whether anyone in the household is vulnerable;
- made sure that written contact with you is in plain English and that it includes details of sources of help, such as Citizens Advice consumer service;
- made a number of attempts to contact you using different methods such as telephone and email and at different times of the day;
- visited the property and looked for any visual signs indicating vulnerability;
- checked whether the property is unoccupied, either on a temporary or permanent basis;
- considered whether the occupancy of the property has changed.

Last-minute negotiations

If the supplier agrees not to disconnect at the last minute, but an official turns up to carry out the disconnection, the disconnection should not be agreed to, and the official should be asked to contact the supplier's office. Many suppliers accept payment on the doorstep, but some make an extra charge to cover their expenses. Always get a receipt to establish payment.

Some electricity suppliers' disconnection officials routinely carry prepayment meters with them and will offer you one as an alternative, even at this late stage. If you accept the meter, check that it has been set to collect arrears at a rate you can afford. If it has not, ask the supplier to change the setting. Do not be put off by such statements as 'it cannot be changed' or 'it is set at the factory'. This is not the case. It is unlikely that a gas supplier would try to fit a prepayment meter straight away – the system has to be purged and re-lit first.

If you refuse to allow entry, the supplier has to obtain a warrant or may disconnect from the road. This costs more, unless you are able to negotiate keeping your supply and paying off the arrears at a rate you can afford in the meantime (see Chapter 7).

---

5. **Getting your supply reconnected**

The supplier must reconnect the supply within 24 hours if:[30]
- you pay your outstanding bill, together with the expenses of disconnection and reconnection and any security deposit; *or*
- you reach an agreement with the supplier to pay off the arrears in instalments as a condition of being reconnected.

Where you pay or reach an agreement outside working hours, the 24 hours begins at the start of the next working day. If the supplier does not reconnect you within the 24 hours, you are entitled to a £30 payment, unless there is an exception (see p83).[31]

In practice, if you agree to accept a prepayment meter set to collect the arrears, the supplier will reconnect your supply. You may have to pay the expenses of disconnection and reconnection separately, but usually they are added to your arrears and collected in instalments through the meter.

If you do not want your supply reconnected

The supplier continues to submit bills regardless of whether or not you want your supply reconnected. If you do not pay, it may seek recovery of the debt through the small claims court or sell on or assign the debt to a third party (such as a debt collecting agency). In practice, few suppliers or debt collectors try to recover through the courts, preferring to send letters in the hope of payment. The reason is that in a small claims court action, legal costs cannot normally be recovered. As a result it would cost more to take a case to court than would be recovered, particularly if you have no disposable income, assets, savings or valuable property.

In some cases, the debt collector may actually be outside the jurisdiction of the county court in England and Wales, being based in Scotland. In such a case, it may be uneconomic for the debt collection company to try to recover the arrears as it is expensive to begin court action.

Assuming you agree you are liable for the bill, try to negotiate payment in instalments prior to any court action. Otherwise, if the supplier has issued a claim in the county court, respond to the claim by completing the statement of your financial circumstances – and ask to pay in instalments. In these circumstances, the court usually orders payment in instalments. If the court has already made an order for payment of the whole debt at once, you could apply to the court to have the order varied to payment by instalments.

There is a fee of £50 for this application, unless you are exempt on grounds of low income. The court should not order you to pay an amount you cannot afford – even if this means you can only afford £1 or £2 a month. You are not liable for the supplier's legal costs if you lost in the small claims court, which is why suppliers seldom commence recovery proceedings for sums under £10,000 (see Chapter 14). The supplier is only able to claim for its fee in starting proceedings. The fees are:

Sums up to £300	£25
Sums from £300-500	£55
Sums from £500-1,000	£80
Sums from £1,000-1,500	£115
Sums from £1,500-3,000	£170
More than £3,000	£335

In Scotland, you can apply for a 'time to pay' direction before a decree (court order). If you break this arrangement by allowing three instalments to pass unpaid, you lose the right to pay by instalments. If you allow a decree to pass without defending or seeking time to pay, you have to wait until the supplier seeks to enforce the decree before you can ask for a 'time to pay' order. You may be liable for the supplier's costs if you lose your case, but only if the debt is over £200. If the value is between £200 and £1,500, the maximum amount awarded to the successful party is £150. If the value is between £1,500 and £3,000 the amount awarded is 10 per cent of the value of the claim.

The costs of disconnection and reconnection

Charges for disconnection and reconnection must be 'reasonable' and must reflect the actual costs involved. Charges vary between suppliers. You will need to contact your supplier to find out how much it charges for disconnection and reconnection.

Ultimately, the question of what is a reasonable cost may be determined by a court. Citizens Advice consumer service or Ofgem may also be able to examine charges. Operators may drop costs at their discretion, and in some cases (where costs appear to be excessive) it may be argued that a supplier is under a duty to mitigate its loss. Legal advice should be sought.

6. **Disputes: unlawful disconnection**

If you dispute that the supplier is entitled to disconnect, you can ask Citizens Advice consumer service to intervene. It can order the supplier to reconnect or continue your supply pending a decision on your dispute.

Unlawful disconnection is an enforcement matter. Suppliers can be forced to comply with the law by Ofgem. If the supplier ignores the order, you can apply for a court order to enforce it (see Appendix 6).

If disconnection was unlawful, you do not have to pay the costs of disconnection or reconnection. You may also have a claim for compensation, for any costs you have incurred as a result of the unlawful disconnection along with any other damage (including to reputation) that may arise from the way in which a supplier acted when disconnecting your supply. These claims may be pursued through the civil courts (see Chapter 14).

For example, a supplier may wrongly disconnect a supply, arising from a mistake by a contractor who visits a household. This can be a problem in multi-occupation buildings or where a contractor suspects the occupants are squatters. In such a case, immediately contact the supplier and be prepared to back up a demand for reconnection with an action through the courts (see Chapter 14). In such a case, an injunction may be sought to order reconnection and a claim for damages included as a result of nuisance and losses caused by being without a supply. In an emergency case, an injunction can be sought outside normal court hours.

Prepayment meters and arrears

In theory, it should not be possible to get into arrears by using a prepayment meter. But in practice, arrears may be transferred from a previous supplier when you switch supplier or they may be set on a prepayment meter if you accept a prepayment meter as an alternative to disconnection.

A problem associated with prepayment meters is that households may disconnect themselves simply by failing to top up the meter through lack of money. If you think this may happen to you, consider using one of the other methods of making payments such as Fuel Direct. Ensure that you provide your supplier with as much information as you can about your particular circumstances.

If you feel that a prepayment meter is being imposed upon you, seek help from Citizens Advice consumer service.

If you have arrears from a previous supply, you may be faced with bills from both your current supplier and your previous supplier. In such a situation, you should pay for your current energy consumption first before paying any arrears to your former supplier, to avoid building up arrears on your ongoing account. Treat the arrears on your old account as a non-priority debt (see Chapter 7). If you have any disposable income after addressing your priority expenditure such as rent, council tax and current fuel supply, then negotiate an affordable repayment plan with your former supplier just as you would with any other unsecured creditor.

Notes

1. **When you can be disconnected for arrears**
   1 Ofgem, *Infographic: energy company performance,* 1 September 2016
   2 Department of Trade and Industry Fifth Report, House of Commons 1 February 2005
   3 Ofgem, *Domestic Suppliers' Social Obligations: 2014 annual report,* 8 September 2015
   4 Energy UK, *The Energy UK Safety Net: Protecting Vulnerable Customers from Disconnection,* February 2016; Ofgem, *Infographic: energy company performance,* 1 September 2016
   5 Condition 26.6 SLC
   6 Sch 6 paras 1 and 2 EA 1989; Sch 4 UA 2000
   7 Sch 6 paras 2(3) EA 1989; Sch 4 UA 2000
   8 Sch 6 paras 2(2)(b) EA 1989; Sch 4 UA 2000
   9 Sch 6 paras 2(2)(a) EA 1989; Sch 4 UA 2000
   10 Sch 6 paras 2 EA 1989; Sch 4 UA 2000
   11 Sch 6 para 27 EA 1989; Sch 4 UA 2000
   12 Sch 6 para 1(6)(a) EA 1989
   13 Sch 2B para 7(1)(b) and (3) GA 1986
   14 Sch 2B para 7(1) and (3) GA 1986
   15 Sch 2B para 7(1) and (3) GA 1986
   16 Sch 2B para 7(5) GA 1986
   17 Sch 2B para 7(4) GA 1986
   18 Sch 2B para 7(3) GA 1986
   19 Sch 2B para 7(5) GA 1986
   20 Condition 27.6(i) SLC
   21 See Conditions 27.5 and 27.6 SLC

2. **Protection from disconnection**
   22 Condition 27.10 SLC
   23 Condition 12A 11(d) SLC
   24 Condition 27 11 SLC
   25 Condition 1 (Definitions) SLC

4. **At the point of disconnection**
   26 Condition 27 SLC
   27 Sch 6 para 2(1)(b) EA; Sch 2B para 73(b) GA 1986
   28 *Associated Provincial Picture Houses v Wednesbury Corporation* [1948] 1 KB 223, per Lord Greene

5. **Getting your supply reconnected**
   30 Reg 6 EG(SP)S Regs
   31 Reg 8 EG(SP)S Regs

Chapter 9

Theft and tampering

This chapter covers:

---

1. **Introduction**

Unlike most other goods, gas and electricity are delivered to you without the supplier being present. Suppliers cannot see what you are doing with their meter or with the fuel they have supplied. This makes suppliers vulnerable to theft. Perhaps because of this vulnerability, suppliers sometimes make allegations of theft on quite flimsy evidence. The consequences of such allegations against you can be severe, as a supplier has the power to disconnect your supply without having to go to court to prove the allegations first.

Theft from, or tampering with, a meter are criminal offences and can result in both criminal prosecution and civil proceedings to recover the alleged debt. However, it is important to realise that not in every case of alleged tampering will you necessarily have to pay for damage or alleged stolen fuel.

A full discussion of criminal law and practice is outside the scope of this book. If you might be liable for prosecution, seek specialist legal advice on your position and how to respond.

If you are legally liable to pay for any loss caused by theft or tampering, a supplier or transporter may be entitled to disconnect the supply, although this right does not follow automatically from liability (see p134).

Ofgem considers theft and tampering to be a serious problem – not least because it estimates that each customer has to pay an additional £6 a year as a result of gas theft alone.[1]

There are a number of regulatory arrangements in place which require suppliers to detect, investigate and prevent theft of electricity and gas.[2] While doing so, suppliers must still ensure that they treat customers fairly and take into account vulnerability. By '**vulnerability**', the regulations mean customers who are of pensionable age, disabled or chronically sick or if the customers will have difficulty in paying all or part of the charges resulting from any theft of fuel.[3] Gas suppliers have implemented a centralised theft risk assessment service to tackle this issue. Ofgem has directed electricity suppliers to implement a centralised theft risk assessment service from 29 February 2016.[4]

Meter ownership

The theft of gas or electricity or interference with a meter often causes damage to the meter itself. The supplier of the gas or electricity and the owner of the meter may not be the same company. Nearly all gas meters are owned by National Grid in its role as the main public gas transporter and electricity meters are owned by the privatised electricity suppliers in their role as distributors of electricity. If you are disconnected for theft or tampering, it will normally be one of these companies that actually carries out the disconnection even if you have a contract with a different supplier for the actual supply of gas or electricity.

2. **Tampering with a meter**

Evidence of tampering

As a supply of gas or electricity is charged for on the basis of metered consumption, the most obvious unlawful method of reducing a potential fuel bill is to interfere with a meter to prevent it registering or to reduce the amount it has registered.

It is a criminal offence to interfere with an electricity meter, punishable with a fine of up to £1,000.[5] An offence is committed where you alter the register of a meter or prevent the meter from duly registering the quantity of electricity supplied.

The offence of abstracting electricity is committed where you dishonestly use electricity without due authority or dishonestly cause electricity to be wasted or diverted.[6] The offence may be tried in the magistrates' court or the Crown Court. On conviction in the magistrates' court, you can be fined or imprisoned for up six months;[7] while on conviction in the Crown Court, the maximum penalty is five years' imprisonment. This penalty would be reserved for the most serious of cases.[8]

There are sometimes tell-tale signs on a meter that has been tampered with – eg, the seals are cut or missing, the casing is cracked or badly scratched or a small hole has been drilled in the side. These descriptions are included not as a guide to people who might want to try it, but for advisers who may have no idea what tampering involves and may need to establish if a meter has been interfered with.

Never assume that an allegation of tampering is correct, whatever technical evidence is quoted by the supplier. The evidence is not always clear-cut and the supplier's experts do not always get it right. You can get your own expert (look for an 'electrical engineer', 'gas engineer' or 'gas installer') to examine the meter for an objective assessment.

Meter tampering is not always seen when the meter is read in the normal way. All meter readers should be trained in detection, but they are only there for a short time. Holes or cracks may be on the far side of the meter in a dark cupboard and, therefore, difficult to spot. In some cases, the dividing up of properties into separate self-contained dwellings and the resultant variations as to the addresses of occupiers can give rise to problems.

Theft due to tampering can be detected by unusual patterns of consumption – eg, if the bills suddenly go down or if they go up after the installation of a new meter. A meter examiner will then come to look at the meter, normally accompanied by a colleague (see Chapter 10) and sometimes by the police. If signs of tampering are detected, the meter will be removed and the supply disconnected. If no evidence is detected there and then, the meter may be taken away for further examination, but a replacement should be left so that the supply is not disconnected straight away. Some electricity suppliers are prepared to install another meter immediately, usually a prepayment meter. However, when a gas supply is disconnected, the system must be purged and re-lit before the supply of gas can be re-started. The people who actually carry out the disconnection are unlikely to have both the expertise and the authority to do this.

As smart meters become more widespread, the scope for unlawful tampering with meters will be reduced as the meter can be controlled remotely and the level of fuel consumption can be regulated without the supplier having to physically remove the meter. Although physical tampering will certainly be more problematic, the question remains whether it will be possible to hack smart meter digital technology in the same way that, for example, home computers are hacked. The householder might hack the device so that usage is under-reported, in order to pay less.

The consensus among consumer groups is that safeguards are needed to manage the collection of consumers' data uploaded by meters and the right to privacy. A third party might hack the device to obtain information about activities within the premises – eg, criminals might be able to use data about fuel consumption to establish when the property is unoccupied. The government publication *Smart meters: a guide* focuses on, among other areas, consumer protections and consumer privacy. This guide can be found at www.gov.uk/guidance/smart-meters-how-they-work. Consumers should be able to choose who can access information collected by the meter by opting out of particular kinds of data management. Additionally, Energy UK has published a data guide for smart meters. This can be found at www.energy-uk.org.uk/policy/smart-meters.html and explains what information meters can collect, who the information may be passed on to and in what circumstances. For example, other industry parties in connection with supply and distribution issues, or police and law enforcement agencies in order to prevent and investigate fraud. Restrictions on the collection of electricity data are also contained in Standard Licence Condition 47 for electricity.

Examining the evidence and the law

If a supplier alleges that a meter has been tampered with, there are three key areas to consider.

- You must establish what exactly is alleged. On what basis has the supplier found that the meter has been tampered with? In some cases it will be obvious that the only explanation for damage to a meter is tampering – eg, there is physical evidence that the meter has been deliberately damaged. In other cases there may be alternative explanations – eg, if meter seals are missing, it may be that these are company seals which were never put on, or were removed by the supplier but not replaced, or have been removed by someone else. Any evidence the supplier has should be presented to you so that you can comment. However, take care in any comment you make, as it is unlikely that you will be cautioned about anything you say possibly being used against you in court.
- You should investigate whether the supplier has evidence that it was you, and not someone else, who tampered with the meter. Usually, the supplier will not know who did the tampering. This can make it more difficult for the supplier to take action against you than it would be if it had clear evidence of your involvement. It will help if you can explain why it has been tampered with – eg, if you know that the meter was taken over from a previous occupier who had tampered with it or that it was damaged by builders. However, evidence that the meter was interfered with while in your custody can, depending on the circumstances, be sufficient to convict you for theft of electricity.[9]
- The evidence must identify the date or dates on which the alleged tampering occurred.

If a supplier alleges that you are liable for theft or tampering, find out which legal provision it is relying on. Different considerations apply when dealing with different parts of the law. Normally the supplier will point to particular provisions in the Gas Act, Electricity Act or Utilities Act, as appropriate, but liability can also arise under general common law principles or under your contract with your supplier.

General common law principles

In England and Wales, when you receive or take on a meter, you become 'bailee' of it. A **'bailee'** is under a duty to take 'reasonable care' of bailed property (in this case, a meter).

In Scotland, 'bailment' does not apply, but the concept of **'restitution'** may be used – ie, if you are in possession of goods which do not belong to you, you are under an obligation to look after them until the owner returns for them.

If you intentionally damage a meter, you could be liable to pay compensation to the supplier. The supplier can sue you, or you may be prosecuted for criminal damage, which is an imprisonable offence. Only a small amount of damage may constitute an offence. On conviction, the court can order you to pay compensation to the supplier to remedy the cost of the damage. Questions of vulnerability are considered before any decision to impose liability for loss arises (see p126).[10]

Contractual liability for meter damage

If you are supplied under a contract, your contractual duty is no higher than your duty as a bailee (see p128), but in the past some suppliers have claimed that customers should pay for damage to a meter even though they had done nothing wrong or were not negligent. For example, if a burglar damaged your meter, you would not normally have to pay for the damage under your duty as a bailee unless your negligence allowed the burglar to get into your home. Read your contract carefully because the relevant term contained there may go wider. However, if the term is so wide that it could be regarded as unfair under the Consumer Rights Act 2015, then it should not bind you, and you should refer it to the Financial Conduct Authority so that its fairness can be looked at (see Chapter 14).

Responsibility for meters under the Gas and Electricity Acts

In nearly all cases the meter will have been provided by a gas supplier or transporter, or hired/loaned by an electricity supplier and, under these circumstances, the meter is its responsibility.[11]

If you hire the meter, you may have to enter into a hire agreement with the supplier. Suppliers have no powers to impose conditions about the care of the meter in such an agreement that go further than those allowed under the Acts or their licence conditions.

There are three specific meter offences under the Acts, each punishable by a fine of up to £1,000:[12]

- damaging or allowing damage to any meter, gas fitting or electrical plant or line;
- altering the meter index or register by which consumption is measured;
- preventing the meter from registering properly.

In each case, the offence can only have been committed if the act was done intentionally or a result of culpable negligence – ie, if it was your fault.

If you are prosecuted for either of the latter two offences, possession of artificial means for altering the way the meter is registering will be taken as *prima facie* evidence that the alteration was caused by you.[13] If such artificial means are not found, a conviction would be difficult to obtain, especially if the case concerns a property in multiple occupation or where there has been a burglary.

3. **Theft of fuel**

Theft of gas and dishonest use or 'abstraction' of electricity are criminal offences.[14] Penalties include fines or imprisonment, offences being triable either in the magistrates' court or the Crown Court before a jury. You can be convicted of theft even if there is no damage or evidence of interference with a meter.[15] The offence of **'abstraction'** is committed where there is use of electricity without the authority of the electricity supplier by a person who has no intention of paying for it.[16]

The key to an offence under these offences is the question of dishonesty. If you genuinely believed you were entitled to use the fuel or had paid or would be paying for the fuel concerned, then the elements of the offence cannot be proved. The issue of what is honest or dishonest use is the standard applied by a jury as being the standards of honest, ordinary people.[17] Dishonesty in the context of an offence of abstraction requires knowledge that electricity was being consumed, coupled with an assumption that it would not be paid for. A genuine belief and intention that you will pay for the electricity you have used – even in a case where a person reconnects a supply without the permission of the electricity company – will provide a defence.[18] It is an offence to dishonestly use electricity without due authority, so a defence may exist where a person honestly believed they had been authorised to use the electricity.[19] Thus, if only one person in a multi-occupation household knows about the unlawful consumption but other occupants are ignorant of it, only the person with knowledge may be convicted of the offence. It is improper to infer guilt simply on the basis of a close relationship between members of a household where an offence has occurred.[20] Prosecutions may arise in connection with other offences which involve the improper use of electricity – eg, where electricity is stolen to facilitate the cultivation of drugs. Liability can also be imposed where third parties have been knowingly hired to carry out work such as altering the arrangements for electricity supplies to premises in order to commit criminal offences. Deliberate attempts, even if no electricity or gas is illegally abstracted or loss is incurred by the supplier will also be prosecuted.[21]

In the case of use of gas or electricity by squatters, it is important that the supplier is notified as to the use as soon as practicable and that an undertaking of willingness to pay is given. If a squatter moves into premises, uses electricity or gas and then moves out without the intention of paying for it, an offence is committed. In one case, the police arrested squatters who were in the process of moving out of a property in which lights were on. The Lord Chief Justice Lord Taylor stated:

> The defence was that if a bill had come they would have paid. However, they were moving out when the police arrived. They had not apprised the Electricity Board of their arrival, their departure or their identity. In those circumstances it was open to the jury to find that they were acting dishonestly and to convict.[22]

The supplier is entitled to recover the costs of any fuel which has been stolen from the person liable. However, as with tampering, suppliers are more likely to address the issue of the stolen fuel by other means – ie, by threatening disconnection. Where the company transporting gas to your home (normally National Grid) is different from the actual supplier, the transporter also has the right to recover the value of the stolen gas.[23]

Squatting in residential premises is a criminal offence.[24] This does not change the legal position with existing supply of gas or electricity provided to a person living as a squatter where, for example, a supply has already been arranged. Provided the squatter genuinely intends to pay for the gas and electricity supplied, no offence of theft is committed.

It is also important to note that just because you have your name on a gas or electricity bill it does not establish you as living at a property. You are simply a consumer of fuel under contract at the address but this does not mean you are living in the property, either as a trespasser or otherwise.[25]

Sentencing for abstraction offences

Guidance on the sentencing for extracting electricity offences under section 13 of the Theft Act has been issued by the Sentencing Council.[26] A wide range of punishments may be imposed from an absolute discharge to imprisonment. The maximum penalty in the magistrates' court is six months' imprisonment. The maximum penalty is five years' imprisonment in the Crown Court for the most serious offences, usually involving the commission of other crimes.

Many factors affect the level of punishment imposed by the court – eg, the motive and any underlying reasons which led to the offence. Aggravating factors which may lead to a higher penalty include stealing fuel over a long period, electricity being extracted from the property of another, attempts to conceal or dispose of evidence, the impact on the community and your previous criminal record.[27] Factors which may reduce the penalty which the court imposes may include personal vulnerability, previous good character, being a sole or primary carer for dependent relatives and taking steps to address offending behaviour. Credit will also be given for a plea of guilty.[28] Poverty and necessity are not a defence to stealing fuel.[29]

Most offences are prosecuted through the magistrates' court and the most common sentence is a fine. The Crown Court may set an unlimited fine, but in both the Crown Court and the magistrates' court the level of any fine will be linked to the ability of the person to pay it.[30] Where a fine remains unpaid, it may be enforced as a civil debt recoverable through the court machinery with imprisonment for default where a debtor has the ability to pay but fails to do so. The court may also make a compensation order.[31]

Accuracy of meters and estimates of stolen fuel

In cases of alleged tampering or theft of fuel, suppliers will try to recover the cost of fuel stolen by estimating the consumption during the period of tampering or theft. This often leads to a dispute about whether or not a meter has recorded consumption accurately. Either party can refer the matter for consideration to a meter examiner.

If the consumption of fuel has been under-recorded, whether because of tampering or otherwise, extra charges will be due. Suppliers will claim that, since the meter has been tampered with, consumption must be estimated – and they often come up with extremely high estimates. If you dispute an estimate and want to challenge it, ask the supplier how the estimate was made and on what assumptions. Just because a meter has been tampered with does not necessarily mean that fuel was successfully stolen – it is still up to the supplier to prove that it was.

There are various ways in which suppliers estimate consumption. One measure is to compare your consumption during the period of tampering with your normal rate of consumption, either before the meter was tampered with or after its replacement. The comparison should be over a period of at least a year, to produce an accurate figure, as consumption tends to increase in winter.

This method may not be appropriate in your case because, for example, your pattern of consumption has changed, or you have recently moved home, or because the supplier claims tampering began after the meter was last read or inspected. There is another method, based on the number and type of appliances you use. Suppliers make assumptions about the running costs of appliances and how often you use them, and then calculate the level of consumption in accordance with those assumptions. Look at the findings critically to see if they bear any relationship to actual usage. Suppliers sometimes assume the existence of appliances which you do not actually have, or that you use the appliances you do have for maximum periods of time and at maximum settings.

Also note when the supplier is alleging that any tampering began. Evidence of tampering may be clear, but not the start date or the period during which it took place. For example, the more times the meter has been read, the less likely it is that tampering would not have been noticed by a meter reader, which shortens the period during which the tampering is likely to have started. Under Standard Licence Condition 12A, the supplier must take all reasonable steps to prevent and detect:

- the theft or abstraction of electricity at premises supplied;
- damage to any electrical plant, electric line or metering equipment through which such premises are supplied with electricity; *and*
- interference with any metering equipment through which such premises are supplied with electricity.

As a result, all electricity suppliers must tell the owner of a meter if they spot any signs of tampering.[32]

If a meter examiner has been called in, s/he will decide the amount of extra fuel you should be charged for. You can get your own electrical expert (look for an 'electrical engineer' or consult Ofgem for licensed meter companies) to make an independent assessment for you. See Chapter 14 for other methods of solving disputes about charges.

Inspection of meters

There are two points to make about the inspection of meters which have been allegedly tampered with.

- A gas supplier is supposed to ensure that your meter is inspected at least once every two years, including a check to see if it has been tampered with.[33] If the supplier claims that the meter has been tampered with for more than two years (hence it can try to claim more than two years' worth of stolen gas), you can point out that this suggests it has breached its obligation to inspect the meter. Under standards of performance imposed by Ofgem, electricity meters should be read at least once every two years.

- In cases of alleged tampering or theft of fuel, suppliers often remove meters quickly. Electricity suppliers must keep meters they have removed because of tampering until Ofgem says otherwise.[34] Meter providers state in their relevant code of practice the minimum length of time they will keep a damaged meter. In the event of legal action, you will need to have your own expert inspect the meter, so check that the meter is being retained correctly and, if necessary, quote the code of practice.

Cloned keys and fuel credit cards

Energy UK has indicated that illegal top-up keys were used 88,300 times during 2011. These are sold illegally, often door-to-door, and purport to reduce electricity consumption or alter meter readings. If you use one, not only will you lose the money but your supplier can detect that the energy used has not been bought legally and you will end up paying twice for your fuel. Anyone knowingly using or selling such a key meter could be prosecuted for theft or for other offences. The fuel industry has committed a significant amount of money to developing technology that can identify and disable cloned cards and keys.

You should only buy top-ups from official outlets. Energy supply companies never sell top-ups door-to-door. If you have bought a discounted top-up without realising it is illegal, stop using the key and contact your fuel supplier. Seek advice from Citizens Advice consumer service about the best way to address any arrears which have accrued while using the key unknowingly.

4. **Disconnection of the supply**

If a supplier alleges theft or tampering, as well as holding you liable for damage to the meter or any financial loss, it may also want to disconnect the supply until you make arrangements to pay for the loss. You should seek legal advice if the supplier intends to bring a civil claim against you on the basis of theft if you have not been charged or convicted of any offence in the criminal courts.

Disconnection powers arise under a number of different provisions and it is useful to find out which power the supplier is relying on. Each power has its own limitations and it is important to make sure they are not exceeded. In particular, the powers to disconnect for damage to, or tampering with, a meter are different from the power to disconnect for arrears. The supplier should clarify which power it is exercising when seeking to disconnect a supply. For instance, longer notice must be given before disconnection for arrears takes place, but in tampering cases only 24 hours may be given for gas and no notice at all for electricity, on the basis that the tamperer could be forewarned to get rid of the evidence. See Chapter 10 for the supplier's rights to enter your home in order to carry out the disconnection.

Injunctions

If suppliers exceed their powers (eg, by refusing to reconnect your supply unless you pay excessive charges) you may be able to get a court order (an 'interlocutory injunction') requiring the supplier to reconnect until the dispute is resolved. Sometimes the threat to seek an injunction may be sufficient to persuade a supplier to reconnect.[35] **Note:** applying for an injunction can be costly. You will need to pay a court fee of at least £255 and should also seek advice from a solicitor before using this remedy. Legal Aid is not available for advice about this type of debt, so you would need to meet the cost of instructing a solicitor yourself.

See Chapter 14 for more information.

Specific powers of disconnection

There are three specific meter offences under the Electricity and Gas Acts which are discussed on p129. If any of the offences are committed in respect of a gas meter or fitting, the supplier can only disconnect the supply of the person who has committed the offence. This is also the case with the offence of damaging an electricity meter or electrical line or plant.[36]

However, if anyone tampers with an electricity meter so as to commit one of the other two offences, the supplier can disconnect the supply from the premises regardless of whether the person who committed the offence is the actual customer, and whether or not other users of electricity live there.

Suppliers' rights to enter your home to disconnect are dealt with in Chapter 10. However, it is worth pointing out that gas suppliers and transporters do not have the right to enter to disconnect for theft or tampering unless they have given

24 hours' notice or have a warrant from a magistrate or, in Scotland, a sheriff or Justice of the Peace.

To get a criminal conviction, it must be proved beyond all reasonable doubt that an offence has been committed. The onus is on your supplier to prove that you have committed an offence. However, to exercise its power to discontinue the supply, the supplier needs only to be able to prove it on the balance of probabilities – ie, it is more likely than not. This test is easier to satisfy than that required to prove a criminal offence. It is important to note that there does not have to be an actual conviction before the power to disconnect can be used.

The supplier can only discontinue the supply until the matter has been remedied.[37] In the case of tampering with a meter, this includes paying for the cost of any damage to the meter and for any stolen fuel, but the two should be treated separately. Obviously tampering is normally carried out in order to reduce the fuel bill, which amounts to theft. However, if a meter has been damaged or tampered with, this does not necessarily mean any fuel has successfully been stolen and proof of damage or tampering is no proof that any money is owing in respect of fuel. (For example, tampering can inadvertently result in the meter actually registering a higher consumption.) Therefore, unless the supplier can show that, on the balance of probabilities, the damage in question caused financial loss other than the cost of replacing the meter, the matter will be remedied once that cost has been met.

Charges for stolen fuel

Often the supplier assesses an amount of fuel which it thinks has been stolen, and demands payment for that before it reconnects. It can only do this if it can prove that there was a theft, and that it was caused by the particular damage in question.[38]

If the supplier can prove that fuel has been stolen, and can justify its assessment of its value, then it can disconnect for non-payment. However, suppliers cannot use this power if the amount charged is 'genuinely in dispute' (see Chapter 7).

Even when charging for gas lost as a result of theft, the supplier must take into account whether the occupants of the premises in question are of pensionable age, disabled or chronically sick when deciding how to recover any sum owed. If you fall into one of these categories, your supplier must use disconnection as a last resort for non-payment of charges arising from theft. An instalment plan or a repayment meter should be offered and the supplier must not disconnect premises during the winter.

Also check carefully any charges for disconnection and reconnection. Gas suppliers are limited by the Gas Act to recovering their 'reasonable expenses'. Otherwise, there is nothing that says exactly what charges can be included, but they must be linked directly to the disconnection and the reasons for it. Typical charges include:

- meter replacement – it is possible for tampering to take place without the meter actually being damaged, so do not pay for a meter to be replaced that is capable of being re-used without repairs;
- visits to your home – check that travel costs and the number of visits are reasonable;
- administration costs of calculating fuel used but not paid for – check that charges are reasonable;
- general administration – check that this is not wholly or partly double-counted within some other charge.

Disconnection for safety reasons

Gas transporters and electricity suppliers have powers to disconnect your supply for safety reasons. A tampered meter can be in an unsafe condition (although not always, as with a meter which is simply missing its seals).

Electricity suppliers can disconnect your supply if they are not satisfied that your meter and wiring are set up and used so as to prevent danger and not to interfere with the supplier's system or anybody else's electricity supply.

Gas transporters have similar powers for 'averting danger to life or property', including dealing with escaping gas. The powers to disconnect are accompanied by rights to enter your property to inspect the relevant fittings and carry out such disconnections (see Chapter 10).

Neither gas transporters nor electricity suppliers have to give notice for disconnection in emergencies. Electricity suppliers must send you a written notice as soon as they can, telling you the reason for the disconnection. If you wish to challenge the decision to disconnect your supply, contact Citizens Advice consumer service. It can refer the matter to Ofgem if a decision on the dispute needs to be enforced (see Chapter 14).

Gas transporters must send you a written notice within five days of the disconnection, telling you the nature of the defect, the danger involved and what action has been taken. If you want to object, you have 21 days to appeal to the Secretary of State for Environment, Food and Rural Affairs against the disconnection. The meter stays disconnected until the fault is remedied or the appeal is successful. Reconnection without the consent of the appropriate authorities (ie, the gas transporter or the Secretary of State) is a criminal offence.[39]

When the supply is disconnected for safety reasons, a supplier may provide alternative appliances (eg, electric heaters and cookers) although this is unlikely if tampering is thought to be involved.

Gas suppliers are obliged by their licence conditions to provide a free gas safety check for installations and appliances for some customers up to once a year.[40] The check includes a basic examination and minor work. If any additional work is necessary, there may be a charge. To qualify, you must request the free safety check yourself, and you must:

- be over 60, registered disabled, or receiving a benefit in respect of disability; *and*
- live alone or with a person who also qualifies.

5. **Theft from meters**

Any form of interference with a meter, whether by electronic interference or some method specifically devised to obtain fuel belonging to another without payment, will constitute an offence under the Theft Acts.

Very few prepayment coin meters now remain. However, if your coin meter is broken into, you have two problems:
- convincing the supplier that you were not responsible for the theft; *and*
- the supplier may want you to pay not only for damage to the meter, but also for the stolen contents of the meter.

If you discover a coin meter theft, report it to the police as soon as practicable; this will help rebut allegations that you are responsible.

Suppliers' policies

Regardless of the legal position, suppliers may have policies, including those set out in their codes of practice or staff guidelines, which are more generous than the minimum provisions of the law. Some of these are not published, in order that they cannot be taken advantage of dishonestly, but it is always worth checking the policies of your gas or electricity supplier.

Insurance

Some household insurance policies cover against theft from prepayment meters.

6. **Removal of meters**

Suppliers (and gas transporters and shippers) have powers to remove, inspect and re-install meters.[41] These powers may be exercised when tampering is suspected, as with the other powers discussed in this chapter. Suppliers must install a replacement meter of the same type, and so leave the supply connected on the same terms as before, unless they are exercising powers to disconnect the supply itself. Replacement rather than repair will also occur whenever the internal parts of a meter come to the end of their working life.

The supplier can disconnect the supply by whatever means it thinks fit if it is doing so as a result of non-payment. This includes removing the meter without replacing it. But seven days' notice must be given by a gas supplier, and two days' notice by an electricity supplier. This notice is usually given in the 'final demand'.

An electricity company is also entitled to disconnect a supply and remove a meter even if legal proceedings under the Theft Act are not being pursued. A number of court decisions have held that disconnection could be justified where the supplier could produce, on the balance of probabilities, the civil standard of proof that unlawful abstraction had occurred, even though there may have been no criminal conviction.[42]

Notes

1. Introduction
1 Ofgem, *Tackling gas theft – the way forward,* Consultation, 2012
2 Condition 12A SLC
3 Condition 12A.1(b)(ii) SLC
4 Condition 12A SLC

2. Tampering with a meter
5 Sch 4 para 11 EA 1989
6 s13 TA 1968
7 s32 Magistrates' Courts Act 1980
8 Sentencing Council Guidelines, 1 February 2016, available at www.sentencingcouncil.org.uk/offences/item/abstracting-electricity/
9 *Semple v Hingston* [1992] Greens Weekly Reports 21:1201
10 Condition 12A SLC
11 Sch 7 para 10(2) EA 1989; Sch 2B para 3(3) GA 1986
12 Sch 7 para 11(1) EA 1989; Sch 2B para 10(1)(a) GA 1986
13 Sch 7 para 11(2) EA 1989; Sch 2B para 10(3) GA 1986

3. Theft of fuel
14 s13 (electricity) and s1 (gas) TA 1968; in Scotland, the common law offence of theft
15 *R v McCreadie and Tume* [1992] 96 Cr AppR 143, CA
16 *R v McCreadie and Tume* [1992] 96 Cr AppR 143, CA
17 *Ghosh* (1982) 75 Cr AppR 154

18 *Collins and Fox v Chief Constable of Merseyside* [1988] Crim LR 247; *Boggeln v Williams* [1978] 1 WLR 873
19 s13 TA 1968.
20 *Collins and Fox v Chief Constable of Merseyside* [1988] Crim LR 247
21 Criminal Attempts Act 1981
22 *R v McCreadie and Tume* [1992] 96 Cr AppR 143, CA at 146
23 Sch 2B para 9 GA 1986
24 s144 Legal Aid, Sentencing and Punishment of Offenders Act 2012
25 *Doncaster Borough Council v Stark and Another* [1997] CO/2763/96 5 November 1997, Potts, J; *Frost (Inspector of Taxes) v Feltham* [1981] 1 WLR 452
26 Sentencing Council Guidelines, 1 February 2016
27 Sentencing Council Guidelines, 1 February 2016
28 s144 Criminal Justice Act 2003
29 *Southwark London Borough Council v Williams* [1971] 2 All ER 175
30 *Re Churchill* (No.2) (1967) 1 QB 190
31 Sentencing Council Guidelines, 1 February 2016
32 Condition 12A SLC
33 Condition 12A SLC
34 Condition 12A SLC

4. Disconnection of the supply
35 *Gwenter v Eastern Electricity plc* [1995] *Legal Action,* August 1995, p19
36 Sch 2B para 10(2) GA 1986 Sch 6 para 4(3) EA 1986

37 Condition 12C SLC; Sch 4 UA 2000
38 *R v Director of General of Gas Supply ex
    parte Smith* [1989] (unreported); *R v
    Minister of Energy ex p Guildford* [1998]
    (unreported)
39 ss17(2) and 29 EA 1989; s18(2) GA
    1986; regs 4, 6, 7, 9 and 10 GS(RE)
    Regs; reg 29(4) ES Regs
40 Condition 29 SLC

6. Removal of meters
41 Sch 5 UA 2000
42 *R v Director General of Gas Supply ex parte
    Smith* [1989] QB 31 July (unreported);
    *Director of Gas Supply ex p Sherlock &
    Morris N Ireland* [1996] QB 29
    November (unreported); *R v Seeboard
    PLC & another ex p Robert Guildford*
    [1998] 18 February 1998, per Ognall, J

Chapter 10

· ·

Rights of entry

This chapter covers:
1. Entering your home (below)
2. Right of entry with a warrant (p142)

1. **Entering your home**

The Gas Act 1986 and the Electricity Act 1989 give suppliers and gas transporters certain rights to enter your home. Suppliers do not have any entry rights other than those under the Acts. These rights can only be exercised if:

- you consent; *or*
- the supplier obtains a warrant from a magistrates' court (in Scotland the sheriff court, a Justice of the Peace or a magistrate); *or*
- there is an emergency.

Suppliers emphasise that they will only disconnect your supply as a last resort and where all other measures have failed. However, in practice this is not always the case, and some suppliers may seek a warrant to disconnect ahead of adopting alternative ways of addressing the arrears – eg, by arranging Fuel Direct. Suppliers' licences also contain conditions requiring them to train their representatives and to ensure that they behave appropriately when visiting your home.

Suppliers and their agents should also be aware of paragraphs 70 to 78 of *Taking Control of Goods: National Standards* (see Appendix 4), especially if you are in a vulnerable situation included in the guidance.[1]

These standards are intended for use by all enforcement agents, public and private, and the creditors (in this instance the fuel suppliers) that use their services. This national guidance does not replace local agreements or the legislation, and you should check any codes of practice published by individual suppliers, which may prove helpful. The standards are not legally binding. They can, however, offer a useful benchmark to determine what you can reasonably expect if suppliers instruct agencies to collect debts on their behalf.

Legal powers

Electricity and gas suppliers and gas transporters have the right to enter your home to:[2]

- inspect fittings or to read the meter – no advance notice has to be given;
- disconnect supply on non-payment of bills (this does not apply to gas transporters). Electricity suppliers must give one working day's notice, and gas suppliers 24 hours' notice (this may be waived on grounds of public safety or tampering);
- discontinue supply or remove a meter under their powers in connection with theft and tampering (see Chapter 9). Gas suppliers must give 24 hours' notice;
- discontinue supply or remove a meter where they are no longer wanted. Electricity suppliers must give two working days' notice and gas suppliers 24 hours' notice;
- replace, repair or alter pipes, lines or plant. Electricity suppliers must give five working days' notice (unless it is an emergency, in which case notice must be given as soon as possible afterwards), and gas suppliers seven days' notice.

Notice should be given in writing and can be served by post or by hand, or by attaching it to any obvious part of the premises. Once any required notice has been given, suppliers may use these rights at any reasonable time. 'Reasonable' is not defined but should be taken to mean at reasonable times of the day – ie, not late at night, or on religious festivals and public holidays such as Christmas Day, or when the supplier knows that it would cause you difficulty.

If your supply is disconnected for any reason other than to do with safety, gas suppliers and transporters also have the right to enter your home to check that the gas supply has not been reconnected without consent.

Electricity suppliers do not have the power to inspect or read the meter if you have written to them asking for the supply to be disconnected and this has not been done within a reasonable time.

A gas transporter also has the right to enter your home if it has reasonable cause to suspect that gas is, or might be, escaping, or that escaped gas has entered your premises, in order to do any necessary work to prevent the escape or to avert danger to life or property.

Officials representing the supplier or transporter must produce official identification when using any of the above powers.

If you intentionally obstruct an official exercising any of the above powers of entry, you can be fined up to £1,000, although you cannot be punished if the official does not have a warrant.[3]

Suppliers must leave the premises no less secure than they found them, and must pay compensation for any damage caused.

Always check that the correct person is named in the warrant application and that the fuel debt does not relate to another person – eg, a previous resident.

Licence conditions

Gas and electricity suppliers operate under licences issued by Ofgem (see Chapter 1), which has powers to force the suppliers to keep to the conditions in their licences (see Chapter 14). Licence conditions state that gas and electricity suppliers must send details of their policies on entering customers' homes to Ofgem for approval.

Suppliers' codes of practice require the following.[4]

- Suppliers' representatives visiting or entering your home must be fit and proper persons – eg, they must have no relevant criminal convictions.
- Each representative must be identifiable, including by driving marked vehicles, wearing appropriate clothing and carrying a suitable photocard.
- Each representative must be fully trained about the legal powers discussed above.
- Suppliers must operate password schemes for pensioners, disabled or chronically sick customers. If you want one, you can have a password known only to you and the supplier so that you can identify genuine representatives.
- Representatives must be able to tell you where you can get further help or advice in relation to the supply of gas or electricity.[5]

Check your supplier's website for its code of practice or phone and ask for a copy.

2. **Right of entry with a warrant**

If you do not consent to the supplier entering your premises in accordance with any of the above rights, or there is no adult to give such consent, the supplier or transporter can get a magistrates' warrant (or the Scottish equivalent). In an emergency, a supplier does not need to get a warrant, but can obtain one nevertheless if entry is obstructed despite the emergency. The issue of a warrant is governed by the Rights of Entry (Gas and Electricity Boards) Act 1954.[6] Although the application is dealt with through the magistrates' court, the application is a civil matter, not a criminal matter.

To get the warrant, the supplier must apply to the magistrates' court or, in Scotland, to a Justice of the Peace, a magistrate or a sheriff. The warrant is granted if the court is satisfied that:

- entry to the premises is reasonably required by the supplier;
- the supplier has a right of entry, but that right is subject to getting consent to enter;
- any conditions the supplier is supposed to meet in order to exercise the right of entry (eg, to give notice) have been met.

Also, the court must be satisfied that:

- if the right of entry does not itself have a requirement for notice, 24 hours' notice has been given after which entry was refused; *or*
- there is an emergency and entry has been refused; *or*
- the purpose of entering would be defeated by asking for consent – eg, if tampering is suspected.

A warrant is only valid if it is signed by the Justice of the Peace (or the sheriff in Scotland).

In recent years, the practice has been to use the warrant system for homes to fit prepayment meters rather than actually disconnect a supply. See p99 if you do not want a prepayment meter imposed.

Under the Electricity Act 1989, a warrant of entry remains in force for 28 days. If entry is not sought within this period, the warrant lapses.[7]

A warrant for entry and disconnection may be quashed or set aside by the court where it has been obtained against the wrong person or premises. The supplier must act exactly in accordance with the terms of a warrant. If you are unhappy with an entry by warrant, check its wording precisely. In one case, Offer (the former electricity regulator) decided that a forcible entry to disconnect supply had been illegal because the warrant only covered forcible entry if the supply was left connected.

Notice of application for a warrant

There is no general requirement under the 1954 Act for the supplier or court to inform you that an entry warrant is being applied for, or has been issued. You have no right to be notified or to be present at the hearing. However, your supplier may state in its code of practice that it will inform you. Under the Humans Rights Act 1998 it may be possible to argue that a person affected by the warrant should be notified of the hearing and given an opportunity to make representations to the magistrates' court. This point has yet to be tested, but Article 6 of the European Convention on Human Rights ensures the right to a fair trial and representation in legal proceedings which affect the rights of a person, including civil obligations. The State is under a duty to ensure the effective protection of rights.[8]

Typically a supplier serves a notice informing you that you may attend.

Some letters may state that the police may be in attendance. This is wrong and misleading. The police should not be involved because the warrant is a civil matter, not a criminal one. The police have no powers to enforce the warrant, as it is a private dispute between you and the supplier. Only if there is a threat of violence at the property should there be any involvement by the police, and then only to restrain a breach of the peace such as a fight. A complaint should be made where a letter contains such a suggestion.

Contacting the supplier in advance

Wherever possible, you or your representative should contact the supplier in advance of the hearing as there is still the possibility of negotiation.

In some cases, the supplier may withdraw the application prior to the hearing, particularly if the application may be contested or if you are vulnerable. Reference should be made to *Taking Control of Goods: National Standards* (see Appendix 4) in a case of vulnerability. If a supplier disregards the guidance, a complaint can be made, as well as the matter being brought to the attention of the magistrates' court.

Representations should be made in writing, and may be faxed or emailed direct to the customer relations or the complaints department. Your supplier should provide you with the contact details for the department which deals with complaints. An explanation of the vulnerability and an outline of the issues should be included.

If a defence can be shown, or there are factors that the court should consider, mention these. You should be prepared to provide further details and attend the court hearing.

Disconnection of smart meters without a warrant

Suppliers are bound by the same rules applicable to those on non-smart meters with regard to disconnection of supply. However, as they have remote access to your meter, and do not have to gain physical access to your property to switch off the supply, they do not have to apply for an entry warrant.

Restrictions on disconnection

Standard Licence Condition (SLC) 27 provides that a supplier should not disconnect in winter (see p117) and that disconnection should be a last resort. When contacting the supplier or its agent, mention these restrictions, and at court where appropriate.

The approach of the court

The court should not grant the warrant unless it is satisfied that the legal requirements have been met; but in practice, courts tend to rubber-stamp suppliers' applications for warrants in the absence of the customer. However, the Court of Appeal has emphasised the importance of carefully scrutinising warrant applications to gain entry to private homes.[9] Magistrates have discretion under the 1954 Act and are expected to exercise that discretion reasonably in each case, considering all relevant factors, disregarding irrelevant ones and not acting perversely.[10] If you suspect that your supplier will be applying for a warrant, write to them. You should set out the reasons why a warrant should not be granted and send a copy to the court, asking that it be shown to the magistrate (or other Scottish court officer) who will deal with the application. Letters and representations should be addressed to the Justices' Chief Executive stating that you wish to attend the hearing and make representations at any application. Letters may be emailed or faxed to the court in urgent cases.

However, as a result of magistrates' court reorganisation and the recent closure of some courts, administration is not always efficient and it may be necessary to attend the court in advance to seek an adjournment, or alternatively on the day set for the hearing. If attending in advance, ask to speak to the duty legal adviser about the application.

Adjournments

If you cannot attend court for a good reason, you may seek an adjournment. The application should only be made if you genuinely intend to contest the warrant application at a later date. Write a letter to the court making a request to allow an adjournment of the warrant application, to allow you to attend or be represented. In an emergency, if you cannot get to court, you may be able to obtain the adjournment by telephone. If you are in a vulnerable situation or if the disconnection is being investigated by an official body, there may be grounds for an adjournment on these grounds. Reasonable consideration should be given to an application to adjourn.

It is important not to leave the matter to the last minute. Act promptly yourself and, in cases where the energy company has failed to respond to representations, this should be put forward as reasons for the delay.

The court may also be willing to adjourn the hearing where there is a contradiction between the information contained in the summons or notice of hearing for the warrant and the evidence actually produced in court on behalf of the supplier.[11]

On attending court it may be possible to negotiate with the agent representing the supplier, and in some cases it may be possible to have the warrant application withdrawn on the day of the hearing. In some cases, simply attending court and being prepared to contest the warrant leads the supplier to withdraw the application.

You may represent yourself in court, be represented by a lawyer, or may be assisted by a 'McKenzie friend' (see p234).

Proportionality and Human Rights Act principles

The court has discretion whether to grant a warrant and must act reasonably.[12] In considering whether to exercise its discretion to grant a warrant, the court should have regard to human rights principles under European law including the 'doctrine of proportionality' – ie, any legal measures applied against citizens of member states and affecting their rights must be proportional to the ends achieved.

The law has yet to be tested, but it is at least arguable that an application for a warrant to gain entry to disconnect electricity or gas may be a disproportionate measure in the case of a vulnerable household – eg, a lone parent receiving only benefit income. In a case of fine enforcement, the High Court held that enforcement activity may be disproportionate as a measure and contrary to Article 8 of the European Convention on Human Rights (protecting rights to the home and family life).[13]

Applying the principle of proportionality, it would appear open to a magistrates' court to decline to issue a warrant where a debt is relatively small and the hardship caused to a vulnerable household would be severe. A magistrates' court should consider the position of any children residing in the property and anyone with disabilities who may be affected. It is therefore important that a financial statement (see p96) and details of all persons residing in the property are provided to the court at the hearing. In the event that you are too ill to attend, send details to the court in writing in time for the hearing.

Similarly, it has not been determined whether it is correct for an energy company to pursue a warrant where another option may be available – eg, Fuel Direct or a payment plan. You should certainly raise any failure to do so at the hearing, as your supplier has a duty to consider these options under SLC 27.

Defects in the 1954 legislation

From anecdotal evidence and experience, it appears that a number of major suppliers have doubts as to the applicability of the 1954 Act. The legislation dates from the period when energy companies were state-owned and supplies were not provided on the modern contractual basis. The legislation as envisaged in 1954 was not designed to accomplish the instalment of prepayment meters, which amount to a change in the terms and conditions of supply. Therefore, there is an argument that the use of a warrant to fit a prepayment meter is not within the powers granted under the Act as envisaged in 1954, and that you can legitimately object to the change in the terms and conditions. Sometimes energy companies are reluctant to tackle these arguments in court and may withdraw the warrant.

If a warrant application is challenged in court, questions should also be asked about the cost of the application. Some suppliers (or companies acting on their behalf) will add £300 or more for the cost of seeking an individual warrant against one, even though they may be making 10 or more such applications at the same time. The court has discretion with regard to costs. This means that you may ask the magistrates' court to consider costs and whether they are reasonable. The additional costs added could also be referred to Ofgem or to Citizens Advice consumer service for examination as to fairness.

Liability for negligence and improperly obtained warrants

A gas or electricity operator is protected against liability in civil law for acts in accordance with executing the warrant. However, a gas or electricity operator who gains entry under a warrant remains liable for any wrongful acts or defaults committed during the course of executing the warrant against the premises or where a warrant has been obtained in bad faith or for an improper purpose.[14] Therefore, a fuel operator may be liable to action in negligence, trespass or nuisance where the warrant is executed against the wrong premises or where damage is caused in the process of disconnection. If a supplier knowingly tries to force entry without a warrant this is a criminal offence, such as criminal damage. Pecuniary losses which arise for wrongful disconnection may be recovered and can include claims for damage to reputation.[15] A householder who resists such entry is entitled to use reasonable force.[16] A warrant should not be granted for entry to premises where the debt relates to the unpaid bills of a previous occupier where a new occupier has moved in.[17]

Appeals from the magistrates' court

Rights of appeal from the magistrates' court on a point of law lie to the High Court under section 111 of the Magistrates' Courts Act 1980 (known as 'case stated' appeals) and also to the High Court by way of judicial review.[18] Seek legal advice before attempting such an appeal and, if you appeal, notify the supplier that you are doing so.

Any application by 'case stated' must be commenced within 21 days of the decision of the magistrates' court. In an appeal by way of judicial review this must be commenced as soon as possible, and in all cases within three months of the decision. An application may be made to obtain a quashing order to cancel out the warrant where it should not have been made in law, where there was procedural unfairness or the decision to issue the warrant was unreasonable in law.

Judicial review is a two-stage process, first requiring an application for leave. An injunction may be sought at the leave stage to prevent the enforcement of the warrant or, if a warrant has been granted, it may be set aside by the High Court.[19] It is always best to seek legal advice before attempting such an appeal. See Chapter 14 for more information.

Notes

● ●

1. Entering your home

1 Ministry of Justice, *Taking Control of Goods: National Standards*, April 2014
2 Sch 6 paras 6(1) and (2) and 7(1) and (2) EA 1989; Sch 2B paras 16, 17, 23(1), 24, 24(2), 26, 27 and 27(1) GA 1986
3 s1(3) RE(GEB)A 1954
4 Code of Practice on Procedures with Respect to Site Access (Electricity); Arrangements in Respect of Powers of Entry; Authorisation of Officers (Gas)
5 Condition 31(1) SLC

2. Right of entry with a warrant

6 s1(3) RE(GEB)A 1954; this applies to Scotland under s11(7)
7 s101 EA 1989, amending s2 RE(GEB)A 1954
8 ECHR Article 6; *Rommelfanger v Germany* (1989) 62 DR 151 and *Diennert v France* (1996) 21 EHRR 554
9 *O'Keegan v Chief Constable of Merseyside* [2003] 1 WLR 2197
10 *Associated Provincial Picture Houses v Wednesbury Corporation* [1948] 1 KB 223
11 s123(2) Magistrates' Courts Act 1980
12 *Associated Provincial Picture Houses v Wednesbury Corporation* [1948] 1 KB 223
13 *R (on the application of Stokes) v Gwent Magistrates' Court* [2001] JPN 766 EWHC Amin 564
14 *O'Keegan v Chief Constable of Merseyside* [2003] 1 WLR 2197
15 *Say (t/a Corby Café) v British Gas Trading Ltd* [2010] EWHC 3946 (QB)
16 Criminal Damage Act 1971; *Vaughan v McKenzie* [1969] 1 QB 557
17 See *R (on the application of MS Superstore Ltd) v Luton and South Bedfordshire Court* [2013] EWHC 551 (Admin), CO/1046/2011
18 CPR 54
19 See *R(on the application of MS Superstore Ltd) v Luton and South Bedfordshire Court* [2013] EWHC 551 (Admin), CO/1046/2011

Chapter 11

Fuel and benefits

This chapter covers:
1. Benefit entitlement checks (below)
2. Income support, income-based jobseeker's allowance, income-related employment and support allowance, pension credit and universal credit (p150)
3. Cold weather payments and winter fuel payments (p151)
4. Housing benefit and universal credit housing costs element (p154)
5. Impact of charitable payments on benefits (p157)
6. Local welfare assistance schemes (p157)
7. Budgeting loans and budgeting advances (p158)
8. Fuel Direct (p159)

1. Benefit entitlement checks

If you cannot afford to pay for fuel or related expenditure, obtain specialist benefits advice to ensure that you are receiving your full entitlement. Do not delay in this, as there are time limits for claiming all benefits and restricted opportunities for backdating. Do not be put off from seeking advice and do not assume that you are not entitled to any help or to more help than you are getting at present. Whatever your circumstances, your local Citizens Advice Bureau, advice centre or welfare rights service should be able to provide you with a benefits check free of charge. In Wales and Scotland, you can get a free benefit check from the government funded energy efficiency schemes:

* in Wales, Nest: freephone 0808 808 2244;
* in Scotland, Home Energy Scotland: freephone 0808 808 2282.

If you are refused a benefit and need to challenge the decision, consult CPAG's *Welfare Benefits and Tax Credits Handbook* for detailed information and seek advice. If you are not entitled to any benefit, see Chapter 12.

2. Income support, income-based jobseeker's allowance, income-related employment and support allowance, pension credit and universal credit

Income support (IS), income-based jobseeker's allowance (JSA), income-related employment and support allowance (ESA), pension credit (PC) and universal credit (UC) are national 'safety net' means-tested benefits. The benefits are administered by the Department for Work and Pensions and are key benefits if you need direct help with fuel-related costs. If you are on these benefits, you may be able to avoid disconnection by paying for your fuel and any arrears through the Fuel Direct scheme (see p159).

For full details about rules of entitlement and how to calculate benefits, see CPAG's *Welfare Benefits and Tax Credits Handbook*.

This chapter does not apply to people who are subject to immigration control, whose immigration status may be jeopardised if they claim certain benefits. See CPAG's *Benefits for Migrants Handbook* for further details.

Income support

IS is a benefit for people aged 16 or over and under the qualifying age for PC who are on a low income. Only certain groups can get IS – eg, lone parents with a child under five and people caring for a disabled person. You cannot get IS if you are working for 16 hours or more a week (and/or if your partner is working 24 hours or more a week) or if you have savings of over £16,000. Your income must be less than the set amount the law says you need to live on (known as your 'applicable amount').

Income-based jobseeker's allowance

Income-based JSA is a benefit for people aged 18 (or in some cases 16) or over and under pension age who are available for and actively seeking work and are on a low income. You cannot get income-based JSA if you are working for 16 hours or more a week (and/or if your partner is working 24 hours or more a week) or if you have savings of over £16,000. Your income must be less than the set amount the law says you need to live on (known as your 'applicable amount').

Income-related employment and support allowance

Income-related ESA is a benefit for people aged 16 or over and under pension age who are unable to work due to ill health or disability. You cannot get ESA if your partner is working 24 hours or more a week or if you have savings of over £16,000.

Your income must be less than the set amount the law says you need to live on (known as your 'applicable amount').

Pension credit

PC is a benefit for people who have reached the 'qualifying age'. For claimants born before 6 April 1950 this is 60. For claimants born on or after 6 April 1950 it is gradually increasing to 66 by 2020 and will eventually go up to 68. PC has two different elements:

* **guarantee credit**, designed to bring your income up to a certain level; *and*
* **savings credit**, which is intended to 'reward' you for making provision for your retirement above the basic state retirement pension. The savings credit is being phased out from April 2016.

Universal credit

Universal credit (UC) is gradually replacing:

* IS;
* income-based JSA;
* income-related ESA;
* housing benefit;
* working tax credit;
* child tax credit.

UC is for people aged 18 and over (or in some cases 16 and over) and under PC age who are on a low income, whether they are in or out of work. You can claim regardless of your circumstances provided you meet the basic conditions about age, education and residence, and you do not have savings or capital of over £16,000. UC includes amounts for adults, dependent children, disability, caring responsibilities, rent or mortgage interest and childcare costs.

See p154 for UC housing costs and fuel charges.

3. **Cold weather payments and winter fuel payments**

There are two types of payments which provide extra help for fuel during cold periods of weather – cold weather payments and winter fuel payments. They have different eligibility rules.

Note: in future the rules on cold weather payments and winter fuel payments may be different in Scotland.

Cold weather payments

Cold weather payments are payments made to pension credit (PC) claimants and some income support (IS), income-based jobseeker's allowance (JSA), income-related employment and support (ESA) and universal credit (UC) claimants. Cold weather payments are intended to assist with the extra costs of heating when the weather has been exceptionally cold for at least seven consecutive days.

A period of cold weather
This is a period of seven consecutive days during which the average of the mean daily temperature, as forecast or recorded for that period at your designated local weather station, is equal to or below zero degrees celsius.[1]

Who qualifies

You qualify for a cold weather payment if:
- a period of cold weather has been forecast or recorded for the area in which your normal home is situated;[2] *and*
- you have been awarded PC (guarantee or savings credit) for at least one day during the period of cold weather. You also qualify if you have been awarded IS, income-based JSA, income-related ESA or UC for at least one day during the period of cold weather *and*:[3]
  - your IS or income-based JSA includes a disability, severe disability, enhanced disability, disabled child or pensioner premium; *or*
  - your income-related ESA includes the pensioner premium, severe disability premium, enhanced disability premium, or the work-related activity or support component; *or*
  - your UC includes an increase for a disabled or severely disabled child; *or*
  - your UC includes the limited capability for work or limited capability for work-related activity element (or would except that you get the carer element instead) and you are not in employment or gainful self-employment during the period of cold weather or on the day it is forecast; *or*
  - you are responsible for a child under five; *or*
  - you are getting child tax credit which includes a disability or severe disability element; *and*
- you are not living in a care home; *and*
- you are not a person subject to immigration control.

Amount of payment

£25 is paid for each week of cold weather.[4]

Claiming and getting paid

You do not need to make a claim for a cold weather payment. The DWP should automatically pay you if you qualify. If you do not receive payment and think that you may be entitled, contact your local Jobcentre Plus or Pension Centre office. A payment cannot be made more than 26 weeks from the last day of the winter period (1 November to 31 March) in which the cold weather period fell.[5]

Challenging a decision

If you do not receive a cold weather payment to which you think you are entitled, submit a written claim for it and ask for a written decision. If you are refused, you can ask for a revision of the decision. Do so within one month of receiving the decision. If you are still unhappy with the outcome, you can appeal to the First-tier Tribunal.

Winter fuel payments

A winter fuel payment is a yearly tax free payment to help people pay for their heating in the winter. Getting the winter fuel payment does not affect any other benefits you may get.

Who qualifies

You qualify for a winter fuel payment if:[6]

- you have reached the qualifying age for PC in the week beginning on the third Monday in September (the **'qualifying week'**). For winter 2016/17, this means that your date of birth is on or before 5 May 1953 and for winter 2017/18, this means that your date of birth is on or before 5 August 1953; *and*
- you are ordinarily resident in Great Britain. **Note:** you may be entitled if you live in Switzerland or in another European Economic Area country, excluding Cyprus, France, Gibraltar, Greece, Malta, Portugal and Spain; *and*
- you claim in time (if a claim is required – see p154); *and*
- you are not excluded under the rules below.

Exclusions

You are excluded from entitlement if, during the qualifying week (see above):[7]

- you are serving a custodial sentence; *or*
- you have been receiving free inpatient treatment for more than 52 weeks in a hospital or similar institution; *or*
- you are receiving PC, income-based JSA or income-related ESA and you are living in residential care. You count as living in residential care if you are living in a care home throughout the qualifying week and the 12 preceding weeks, disregarding any temporary absences;[8] *or*
- you are a person subject to immigration control (although there are exceptions to this rule).

Amount of payment

These are the rates that apply in 2016/17:

- £200 if you are aged between the qualifying age for PC and 79 (inclusive) in the qualifying week; *or*
- £300 if you are aged 80 or over in the qualifying week.

If you do not get PC, income-based JSA or income-related ESA and you share your accommodation with another qualifying person (whether as a partner or friend), you get £100 if you are both aged between the qualifying age for PC and 79, or £150 if you are both aged 80 or over. If only one of you is aged 80 or over, that person gets £200 and the other person £100. If you do get PC, income-based JSA or income-related ESA, you (and your partner if you have one) get £200 if one or both of you is aged between the qualifying age for PC and 79, or £300 if one or both of you is aged 80 or over, regardless of whether there is anyone else in your household who qualifies.[9]

If you are living in residential care in the qualifying week and are not getting PC, income-based JSA or income-related ESA, you are entitled to a payment of £100 if you are aged between the qualifying age for PC and 79, or £150 if you are aged 80 or over.[10]

Claiming and getting paid

You should automatically receive a payment without having to make a claim if you received a payment the previous year, or you are getting a state retirement pension or any other social security benefit (apart from child benefit or housing benefit) in the qualifying week.[11]

Otherwise, you must claim a winter fuel payment before 31 March following the qualifying week.[12] To ensure you receive your payment before Christmas, submit your claim before the qualifying week. Claim via the winter fuel payment helpline on 0345 915 1515 (textphone 0345 606 0285) or download a claim form from www.gov.uk/winter-fuel-payment.

If you are a member of a couple and your partner is receiving IS, the payment can be made to either of you (even if your partner is under qualifying age for PC).[13]

Payments are usually made between mid-November and Christmas.

Challenging a decision

Decisions can be challenged by revision, supersession, mandatory reconsideration or appeal. To get a decision, you may have to submit a written claim and request a written decision.

4. **Housing benefit and universal credit housing costs element**

Housing benefit (HB) is a means-tested benefit intended to help low-income households with rent payments. You can claim HB whether you are in work or out of work.

Note: for many people, HB is being replaced by universal credit (UC) which can include a housing costs element. See p156 for information about the housing costs element of UC.

Housing benefit and fuel costs

HB does not assist with most fuel costs that you pay with your rent. You are expected to find the money for these charges from any other income you may have, such as benefits or earnings. However, the following charges may be met by HB.

- Service charges for communal areas – as long as they are separately identified from any other charge for fuel used within your accommodation.[14] Communal areas include access areas – eg, halls, stairways and passageways.[15] In sheltered accommodation only, rooms in common use (eg, a TV room or dining room) can also be included.[16]
- Charges for the provision of a heating system (eg, for boiler maintenance), if they are separately identified from any other fuel charge.[17]

How fuel charges are calculated

With the exception of those charges listed above, HB does not cover fuel charges which are included in your rent – eg, heating, hot water, lighting and cooking. If you have this kind of fuel charge included in your rent and the amount of your fuel charge can be identified (eg, in your rent agreement, rent book or letter from your landlord), the amount specified is deducted from the total amount of your rent before your HB is calculated.[18] As a result, HB may not cover the full accommodation charges that you are contractually expected to meet. For example, if your rent is £70 a week and your rent agreement states that this includes £15 for heating, £55 would be counted as rent in assessing your entitlement to HB.

If the local council considers that the amount you pay for fuel is unrealistically low compared with the cost of the fuel provided, or if this charge contains an unknown amount for communal areas, it may instead apply a flat-rate deduction (see below). This does not apply if you are a council tenant as the regulations assume that your fuel charges are as specified in your tenancy agreement.[19]

A flat-rate deduction is made if the amount of fuel charges is not specifically identified as part of your rent.[20]

Flat-rate deductions from housing benefit

If fuel charges are included in your rent, the amount of rent which is eligible for HB is calculated by making flat-rate deductions if your fuel charges:[21]

- are not readily identifiable; or
- are considered to be unrealistically low; or

- contain an unknown amount for communal areas as well as for other fuel costs.

Fuel deductions – weekly deductions for ineligible fuel charges (2016/17 rates)
If you and your family occupy more than one room:

Heating	£28.80
Hot water	£3.35
Lighting	£2.30
Cooking	£3.35
Total of all fuel	**£37.80**

If you and your family occupy one room only:
Heating alone, or heating combined with either hot water or lighting or both   £17.23
Cooking   £3.35

If fuel is supplied for more than one purpose, the appropriate charges are added together. If you are a joint tenant, the deductions are apportioned according to your share of the rent.[22]

The local authority must notify you if it has used flat-rate deductions in calculating your entitlement to benefit. It must also explain that these can be varied if you can produce evidence of the actual or approximate amount of the fuel charge.[23] The flat-rate deductions can be varied accordingly. The *Housing Benefit Guidance Manual* used by local authorities says that the lower rate applies if you occupy one room, even if you may share a kitchen or bathroom.[24] Argue for the lower rate deduction if you are forced to occupy one room due to disrepair, damp or mould growth in your home.

Universal credit housing costs element and fuel costs

UC can include a housing costs element for rent or mortgage interest payments. The housing costs element can also cover certain service charges if they are:[25]
- for the cost of services or facilities which are for the use or benefit of people occupying the accommodation; *or*
- fairly attributable to the costs of services or facilities connected with the accommodation that are available for the use or benefit of people occupying the accommodation.

If you live in the social-rented sector (eg, local authority or housing association) or you are an owner-occupier, service charges can be included provided that payment of the charge is a condition of your right to occupy your home, and the charge is for certain types of specified service. This includes fuel costs for communal areas, but not fuel costs relating to your own home.[26]

If you live in the private-rented sector, service charges for fuel costs can be included in the rent, but your housing costs element is restricted by the local housing allowance, based on the market rents in your area.[27]

Discretionary housing payments

The local authority which pays your HB can also make a discretionary housing payment (DHP) if you are entitled to HB or the housing costs element of UC (for rent) and you require additional financial assistance with your housing costs. The onus is on you to apply and the discretion is with the local authority as to whether a payment is appropriate. DHPs cannot be used to cover charges which are excluded from HB or the housing costs element of UC – eg, most fuel charges.

Request an application form from the local authority benefits department.

DHPs are often made on a weekly basis, to 'top up' HB/UC entitlement. It is also possible to receive a DHP as a lump sum – eg, to cover arrears of rent.[28]

There is no right of appeal against a DHP decision, but you can ask the local authority to review its decision. If a local authority acts unreasonably regarding making a decision, it can be subject to judicial review or you can make a complaint to the Local Government Ombudsman (see Chapter 14).

5. Impact of charitable payments on benefits

Charities may sometimes step in to help with fuel or reconnection costs, particularly when 'vulnerable' people have been disconnected. Many Citizens Advice Bureaux and other advice agencies can help with applications for charitable payments.

Most charitable or voluntary payments which are paid regularly are disregarded as income for means-tested benefits. A 'one-off' or occasional payment may count as capital, but is very unlikely to affect benefit entitlement unless it takes your capital above the limit for the benefit you are on.

For detailed information about the effect on benefits of regular and irregular payments for fuel, and the treatment of payments as income or capital, see CPAG's *Welfare Benefits and Tax Credits Handbook*.

6. Local welfare assistance schemes

Each local authority in England and the devolved administrations in Scotland and Wales have local welfare assistance schemes. These schemes may be able to help you with fuel-related costs such as connection and installation charges, draught-proofing and heaters. To find out about the scheme where you live, go to CPAG's local welfare assistance schemes finder at www.cpag.org.uk/lwas.

7. **Budgeting loans and budgeting advances**

Social fund budgeting loans

A budgeting loan is repayable, usually by deduction from ongoing benefit entitlement. The minimum amount of loan is £100 and the maximum is £1,500. You cannot get a budgeting loan if you come under the universal credit (UC) system. Instead, you have to apply for a budgeting advance of UC (see below). You may get a budgeting loan if:

- you are getting income support (IS), income-based jobseeker's allowance (JSA), income-related employment and support allowance (ESA) or pension credit (PC) and have been getting one of these benefits for the past 26 weeks (disregarding one or more breaks of 28 days or less);
- you are not involved in a trade dispute;
- you do not have too much capital (the limit is £1,000, or £2,000 if you or your partner are aged 61 or over).

The budgeting loan must be for one or more of the following:
- furniture and household equipment;
- clothing and footwear;
- rent in advance and/or removal expenses;
- improvement, maintenance and security of the home;
- travelling expenses;
- expenses associated with seeking or re-entering work;
- hire purchase and other debts for any of the above items.

Universal credit budgeting advances

A budgeting advance of UC is repayable. The minimum amount of advance is £100 and the maximum is £348 if you are single with no dependent children, £464 if you are a couple with no dependent children and £812 if you have a dependent child or children (whether you are single or in a couple). You may be able to get a budgeting advance of UC if you require assistance with an 'intermittent expense' and:

- you are getting UC and (unless the expense is necessarily related to employment) have been getting UC, IS, income-based JSA, income-related ESA or PC for the past six months;
- your earnings are below a specified level (in 2016/17, not exceeding £2,600 in the previous six months if you are single, or £3,600 if you are in a couple);
- you do not have any outstanding budgeting advance still to repay and the Department for Work and Pensions is satisfied that you can repay an advance.

8. **Fuel Direct**

The Fuel Direct scheme allows an amount to be deducted from your benefit entitlement and paid directly to your energy supplier. To go onto the Fuel Direct scheme you must be in debt for mains gas or mains electricity and in receipt of income support (IS), income-based jobseeker's allowance (JSA), income-related employment and support allowance (ESA) or pension credit (PC). Deductions can be made from contribution-based JSA or contributory ESA if you have an 'underlying entitlement' to the means-tested version of the benefit – ie, if you were not receiving the contribution-based type of benefit you would instead be getting the means-tested type at the same rate.[29]

See p162 if you are on universal credit (UC).

For a Fuel Direct arrangement to be set up, the Department for Work and Pensions (DWP) and your energy supplier both have to agree. The operation of Fuel Direct involves direct deduction from benefit for both current consumption and for debt recovery. The debt recovery rate is set at £3.70 (during 2016/17) a week, or £7.40 a week if you have debts for both gas and electricity.

Fuel Direct is also known as the DWP's 'third party deduction system'.

Deductions can be made from your benefit if:[30]

- your arrears for gas or electricity are greater than the IS rate for a single person aged 25 and above (£73.10 during 2016/17); *and*
- you continue to need a fuel supply; *and*
- it is in your, or your family's, best interests for direct payments to be made.

You are normally refused if:

- the above do not apply; *or*
- you already have a prepayment meter which has been reset to collect arrears (though if you have a prepayment meter for current consumption only, you could still have the arrears paid by Fuel Direct); *or*
- your supplier does not agree to you paying this way.

How to arrange Fuel Direct

Contact the DWP and your supplier if you want to pay by Fuel Direct. The decision to include you in the Fuel Direct scheme is made by a DWP decision maker. The DWP then contacts your supplier to check that it agrees to you paying in this way and gets the figure to pay for your current consumption.

Before agreeing to deductions, check that you are the person liable for the bill (see Chapter 5).

If the amount of the deduction totals more than 25 per cent of your IS/JSA/ESA applicable amount or PC minimum guarantee (before housing costs), you must give your consent for the deduction to take place. If you receive child tax credit (CTC), this calculation is 25 per cent of the total of your CTC, child benefit and IS/JSA/ESA applicable amount or PC minimum guarantee (before housing costs).[31]

If disconnection is being threatened, let your supplier know that you want to arrange or are arranging Fuel Direct. Suppliers normally delay disconnection for a limited period if they know you are trying to do this.[32] Continue to stay in regular contact with your supplier, informing it about the progress of your application. Take the name and extension number of the person arranging Fuel Direct for you and always keep a note of when you called. It helps if you keep copies of letters/ forms in the event of difficulties. If there are any delays, the supplier can be asked to delay disconnection for a longer period. Check with the DWP that your application has been received and is being dealt with. If disconnection is imminent, ask the DWP to phone the supplier to confirm that Fuel Direct is being arranged and that written confirmation will follow.

Deductions are made at the DWP office before you receive your benefit. If you disagree with any decision about deductions, you can appeal.

Deductions

Deductions can be made to cover arrears, or just to cover weekly costs after the debt has been cleared, or both.[33]

For arrears

- The maximum statutory deduction that can be made for electricity or gas is £3.70 for each fuel debt.
- There is a total maximum deduction of £7.40 altogether for gas and electricity arrears.[34]

For current consumption

- Your supplier gives the DWP an estimate of your weekly consumption. This is usually calculated by looking at your consumption over the past year. It may be appropriate to look at alternative periods of time when calculating your estimated weekly consumption. If the amounts suggested seem high, ask for an explanation – errors are not uncommon. Check that your supplier is not relying on estimated figures.
- The final decision on the amounts deducted rests with the DWP decision maker, who is not bound to accept the supplier's estimates of your current consumption.[35]
- The maximum deduction that can be made without your consent is 25 per cent of your IS, JSA or ESA 'applicable amount' or PC 'minimum guarantee' (your entitlement before any deductions for income, etc. are made). If you get CTC this calculation is 25 per cent of the total of your CTC, child benefit and IS/JSA/ESA applicable amount or PC minimum guarantee (before housing costs). This includes the combined amount of the deductions for arrears and current consumption.[36]
- If you disagree with the amounts proposed by the supplier, ask the decision maker for a different deduction. You will need to provide information about

the assumptions in your own calculation. You will need to conduct your negotiations with care to ensure that the decision maker or the supplier does not assume that you are refusing to join the Fuel Direct scheme, as this could ultimately lead either to disconnection or to the imposition of a prepayment meter. If the decision maker does not agree to accept your calculations, you could always accept the supplier's calculation to ensure entry to the scheme, and then challenge the decision.

Changing the amount of the deduction

The deduction for arrears is a fixed amount and cannot be varied. The figures for your current consumption are normally reviewed regularly.

It is important to make sure that deductions are based on an actual reading of your meter. If the supplier bases your estimated future consumption on estimated readings of your meter, the amount of your deductions is likely to be wrong.

Ask the decision maker to review the amount of deduction if you think the amount is too much or not enough.[37] The decision maker can ask the supplier to provide details of the basis of the supplier's calculation of estimated current consumption.

You may wish to request a review if you can provide evidence that your actual consumption is likely to be different. You may want to do this if:

- an actual meter reading shows that the supplier's calculation is based on wrong information;
- your consumption has increased or decreased because of a change in your circumstances (eg, the birth of a baby, a child leaving home, staying at home more because of illness), a change in a heating system, energy efficiency improvements to your home (eg, insulation, double glazing, draught proofing) or changes to the way in which you use fuel as a result of energy advice.

You can request a review at any time a relevant change occurs. It is sensible to obtain the supplier's agreement to this, as this will make the decision-making process smoother.

Other direct deductions for debts

Other debts can also be paid by direct payments from benefit, and payment of these may be in competition with payments for fuel. The number of deductions for arrears is limited to a maximum of three. Regulations provide that debts are paid in the following order of priority:[38]

- housing costs not covered by the mortgage payment scheme;
- rent arrears (and related charges);
- gas and electricity charges;
- water charges;
- council tax arrears;
- unpaid fines, costs and compensation orders;

- payments for the maintenance of children;
- repayment of integration loans;
- repayment of eligible loans;
- repayment of tax credit overpayments and self-assessment tax debts.

If you have both gas and electricity arrears, the decision maker decides which debt takes priority.

The following debts can also be paid from benefit, but are not mentioned in the regulation governing priority between debts. You should argue that payments for gas or electricity take a higher priority than these debts:

- deductions for overpayments of benefit (but check that the overpayment is recoverable – seek advice if necessary);
- loans from the discretionary social fund.

Change of circumstances

You have a duty to advise the DWP of any changes in your circumstances. This is a normal requirement for benefit claims, but is particularly important with respect to changes in energy supplier. If you switch your gas or electricity supplier, you must immediately notify the DWP.

Sanctions

Benefit sanctions may affect the operation of Fuel Direct, reducing the amount of money from which it is possible to make a deduction. Benefit sanctions are potentially open to challenge by way of revision and then appeal, so seek advice if you are sanctioned.

Universal credit

In some circumstances, deductions can be made from UC for debt and ongoing consumption of mains gas and electricity. This includes debts relating to reconnection of mains gas supply and disconnection or reconnection of mains electricity supply including payment under the green deal plan.[39]

UC is calculated and paid on a monthly basis. The monthly amount that can be deducted in respect of a fuel debt depends on your age and on whether you are single or a couple.

Monthly deductions for fuel debts (2016/17)[40]

Single, under 25	£12.59
Single, 25 or over	£15.89
Couple, both aged under 25	£19.76
Couple, one or both aged 25 or over	£24.94

These amounts are based on 5 per cent of the standard allowance for your circumstances.

In addition, an amount can also be deducted for ongoing fuel consumption, unless this is not required – eg, if you already have a prepayment meter. If a deduction is made for ongoing fuel consumption as well as fuel debt, the maximum amount that can be deducted for both the debt and ongoing consumption is 25 per cent of your standard allowance and any child element, unless you give your consent to a higher deduction being made.[41]

Deductions from UC for current consumption cannot continue once the debt is cleared.[42]

Deductions can be made from UC for other debts, such as rent arrears, housing costs and water charges. A maximum of three deductions can be made at any time.[43]

Where you are working or self-employed, earning over a certain level and getting UC, you are not able to have a deduction for fuel debt made from UC.[44]

Notes

3. Cold weather payments and winter fuel payments
1 Reg 1(2) SFCWP Regs
2 Reg 2(1) and (2) SFCWP Regs
3 Reg 1A(2) and (3) SFCWP Regs
4 Reg 3 SFCWP Regs
5 Reg 2(6) SFCWP Regs
6 Reg 2 SFWFP Regs
7 Regs 2 and 3 SFWFP Regs
8 Reg 1(2) and (3) SFWFP Regs
9 Reg 2 SFWFP Regs
10 Reg 2(2)(b) SFWFP Regs
11 Reg 4 SFWFP Regs
12 Reg 3(1)(b) SFWFP Regs
13 Reg 36(2) SS(C&P) Regs

4. Housing benefit and universal credit housing costs element
14 Sch 1 paras 5 and 6(1)(b) HB Regs; Sch 1 paras 5 and 6(1)(b) HB(SPC) Regs
15 Sch 1 para 8 HB Regs; Sch 1 para 8 HB(SPC) Regs
16 Sch 1 para 8 HB Regs; Sch 1 para 8 HB(SPC) Regs
17 Sch 1 para 8 HB Regs; Sch 1 para 8 HB(SPC) Regs
18 Sch 1 para 6(1) HB Regs; Sch 1 para 6(1) HB(SPC) Regs

19 Sch 1 para 6(1)(a) HB Regs; Sch 1 para 6(1)(a) HB(SPC) Regs
20 Sch 1 para 6(2) HB Regs; Sch 1 para 6(2) HB(SPC) Regs
21 Sch 1 para 6(2) HB Regs; Sch 1 para 6(2) HB(SPC) Regs
22 Reg 12B(4) HB Regs; reg 12B(4) HB(SPC) Regs
23 Sch 1 para 6(4) HB Regs; Sch 1 para 6(4) HB(SPC) Regs
24 para A4 4.912-4.193 HBGM
25 Reg 25(2)(c) and Sch 1 para 7(1), (2) and (4) UC Regs
26 Sch 1 para 8 UC Regs
27 Sch 4 para 3(3) UC Regs
28 *Gargett, R (on the application of) v London Borough of Lambeth* [2008] EWCA Civ 1450

7. Fuel Direct
29 Sch 9 para 1(2) and (3) SS(C&P) Regs
30 Sch 9 para 6(1) SS(C&P) Regs
31 Sch 9 para 8(4) SS(C&P) Regs
32 See your supplier's code of practice for customers who need help with paying their bills
33 Sch 9 para 6(4) SS(C&P) Regs
34 Sch 9 para 6(2) SS(C&P) Regs
35 Sch 9 para 6(4) SS(C&P) Regs

36 Sch 9 para 8 SS(C&P) Regs
37 Sch 9 para 6(4) SS(C&P) Regs
38 Sch 9 para 9 SS(C&P) Regs
39 Sch 6 para 8(2) UC,PIP,JSA&ESA(C&P)
    Regs
40 Sch 6 para 8(4)(a)
    UC,PIP,JSA&ESA(C&P) Regs
41 Sch 6 para 3(3) UC,PIP,JSA&ESA(C&P)
    Regs
42 Sch 6 para 8(2) UC,PIP,JSA&ESA(C&P)
    Regs
43 Sch 6 para 3(1)(b)
    UC,PIP,JSA&ESA(C&P) Regs
44 Sch 6 para 8(5) and
    (6)UC,PIP,JSA&ESA(C&P) Regs

Chapter 12

• •

Energy efficiency and other sources of help

This chapter covers:

---

1. National energy efficiency schemes

Energy efficiency has become an increasingly important part of the government's national energy strategy, which has the shared goals of reducing harmful emissions into the environment and tackling fuel poverty. It is widely accepted that the main cause of fuel poverty in the UK is a combination of low household incomes, high fuel costs and poor energy efficiency. Families with children and working-age households comprise 70 per cent of those in fuel poverty.[1]

Energy efficiency can be defined as use of the minimum amount of energy while maintaining a desired level of a certain activity or service. In other words, energy efficiency is the amount of useful energy output achieved per unit of energy input. Therefore, improving energy efficiency means either achieving more with the same input or achieving the same output with less input.

Often people cannot afford to heat their homes to appropriate levels because they are not sufficiently insulated or in a poor state of repair, or because expensive or inefficient appliances are being used. Substantial savings in fuel bills can be achieved by introducing energy efficiency measures and adopting more energy efficient behaviours.

Successive governments have acknowledged that energy efficiency should play a central role in improving living conditions for the fuel poor in the UK. Official statistics show that the number of fuel poor households in:

• England was estimated at around 2.38 million in 2014, representing approximately 11 per cent of all English households;[2]

- Scotland was estimated at 845,000 in 2014,[3] representing 34.9 per cent of the population;
- Wales is estimated at 291,000 in 2016,[3] or 23 per cent of all Welsh households;[4]
- Northern Ireland was estimated at 294,000 households in 2011,[5] or 42 per cent of households.

An overall UK figure for fuel poverty is no longer measured due to the differences in methodologies, timings and definitions. However, it is estimated by fuel poverty charities that fuel poverty affects over 4 million UK households – this represents approximately 15 per cent of all UK households.[6] There are now two officially accepted ways to define and measure fuel poverty in the UK. According to the original definition, a household is considered to be fuel poor if over 10 per cent of its disposable income is spent on fuel. This definition has been retained in Northern Ireland, Scotland and Wales. However, in England, the 'low income high costs' (LIHC) indicator has been introduced to measure fuel poverty. This is based on the recommendations of the Hills Review.[7] This LIHC methodology is a relative measure of fuel poverty, which also provides a measure of the 'fuel poverty gap' – the reduction in required spending in order for a household not to be considered fuel poor.

The change in definition of fuel poverty for England has in effect reduced the number of fuel poor households by around 2 million, without actually changing the situation of these households.

This chapter looks at what help and support is available, how to access it and who is eligible. See also Chapter 13 for information on how you can exercise your rights against low-standard properties.

Energy Company Obligation

The Energy Company Obligation (ECO) is a UK government energy efficiency scheme designed to help reduce carbon emissions and tackle fuel poverty by placing legal obligations on larger energy suppliers to deliver energy efficiency measures to domestic premises. The current scheme runs until the end of March 2017 – see p167 for future changes.

There are currently three main strands to the ECO:
- Home Heating Cost Reduction Obligation (HHCRO) (see p167) for low-income and vulnerable households;
- Carbon Emissions Reduction Obligation (CERO) (see p168) for those living in hard-to-treat properties;
- Carbon Saving Community Obligation (CSCO) (see p168) for those living in low-income areas and deprived rural areas.

Various measures can be used to satisfy each of the three ECO strands – eg, insulation measures and connections to district heating systems can form part of both the CERO and the CSCO. Under the HHCRO, any measure which reduces the notional cost of heating the property counts towards the target (including boiler replacements or repairs, if aftercare services are provided).

Future changes

In summer 2016, the government consulted about the ECO. It is proposed to extend the ECO for a further year (ie, to have a transition year 2017/18), until the adoption of a new four-year energy efficiency scheme from 2018 to 2022. The 2015 Spending Review set out plans to have a supplier obligation in place until 2022, with a focus on fuel poverty. The details post-March 2017 are therefore as yet only proposals.

Furthermore, the ECO is one of the obligations which will be devolved to the Scottish government under the Scotland Act 2016. Details of what shape this will take have not yet been announced.

Home Heating Cost Reduction Obligation

This is also known as the Affordable Warmth Obligation. This strand of the ECO provides free heating and hot water saving measures (eg, repair or replacement of boilers and electric storage heaters) and insulation (eg, cavity wall insulation) to low-income and vulnerable households.

You are eligible if you are an owner-occupier or live in a private rented property (although you will need the landlord's permission) and a member of your household gets:[8]

- pension credit (PC);
- child tax credit (CTC) and your income is £16, 010 or less;
- working tax credit and your income is £16, 010 or less and:
  - is responsible for a child under 16; *or*
  - gets a disability or severe disability element; *or*
  - is aged 60 years or over;
- income support, income-based jobseeker's allowance or income-related employment and support allowance (and you get the work-related activity or support component) and:
  - is responsible for a child under 16; *or*
  - is responsible for a young person who is 16 or over but under the age of 20 and is in full-time education or approved training; *or*
  - receives one of the following qualifying components:
    - CTC, which includes a disability or severe disability element; *or*
    - a disabled child premium; *or*
    - a disability premium, enhanced disability premium or severe disability premium; *or*
    - a pensioner premium, higher pensioner premium or enhanced pensioner premium;

- universal credit and your income does not exceed £1,250 a month in any assessment period throughout the previous year and:
  - is responsible for a child or qualifying young person; *or*
  - has limited capability for work, or limited capability for work and work-related activity; *or*
  - receives disability living allowance; *or*
  - receives personal independence payment.

Note: check www.gov.uk/energy-company-obligation for the eligibility criteria after March 2017.

Carbon Emissions Reduction Obligation

This is also referred to as the Carbon Saving Obligation. This strand of the ECO provides funding for various energy efficiency measures divided into primary and secondary measures. Examples of primary measures include insulation of solid-walled properties (internal and external wall insulation) and homes with 'hard-to-treat' cavity walls, and connection to district heat systems. In cases where primary measures have been applied, secondary measures such as glazing and draught proofing can be installed. It is not means-tested and can be used in all housing tenures. If you are a tenant, seek the permission of your landlord.

Carbon Saving Community Obligation

This strand provides insulation measures and connection to district heat systems to households in specified areas of low income such as those who live in the lowest 25 per cent of the UK's most poor areas.

Funding can also be given to households in all rural areas. At least 25 per cent of CSCO funding should be used to upgrade more hard-to-reach low-income households in rural communities.

Renewable Heat Incentive

The Renewable Heat Incentive (RHI) is a payment for people living in England, Scotland or Wales for generating heat from renewable sources. RHI payments are intended in part to offset the cost of installing and running a renewables system. Instalment payments are made quarterly for seven years.

You are paid for every kilowatt hour (kWh) of energy produced. The level of payment varies depending on the technology and the system size.

To apply for RHI, you must:
- provide a valid Energy Performance Certificate (EPC) for your property that is less than 24 months old;
- install loft and cavity wall insulation if it is recommended by your EPC;
- have a Microgeneration Certification Scheme (MCS) certificate for your renewable heating installation;
- apply within 12 months of your renewable technology being commissioned.

Eligible technologies are:
• biomass (wood burning) boilers and biomass pellet stoves;
• air source heat pumps;
• ground source heat pumps;
• solar thermal panels.

Only specific makes and models are eligible.[9]

Feed-in tariffs

Under this scheme, energy suppliers must make regular payments to householders and communities in Great Britain who generate their own electricity from renewable or low carbon sources such as solar panels or wind turbines.

The tariffs are paid for every kWh of electricity you generate using a renewable electricity system. They are applicable to households, landlords and businesses. Tariffs can change as often as every three months. Comprehensive tables showing rate changes up to March 2019 are available from Ofgem.[10]

Feed-in tariffs (FITs) provide three benefits.
• You get a payment for electricity produced, even if you use it yourself. Tariffs are paid for up to 20 years and vary depending on the type and scale of the installation. FITs payments are exempt from income tax. The eligibility period starts on the eligibility date and lasts for:
  – solar photovoltaic (PV) – 20 years (25 years for those with an eligibility date before 1 August 2012);
  – wind – 20 years;
  – hydro – 20 years;
  – anaerobic digestion – 20 years;
  – micro-CHP – 10 years.
• You get additional payments for electricity exported to the grid. A payment of up to 4.91p/kWh (correct as of 5 October 2016) is made for any surplus electricity generated and exported to the grid. This is a fixed rate (though there is the option to negotiate an alternative rate with an electricity supplier), regardless of the type of renewable technology. From time to time, the legislation changes and the rates may change as well. Until or unless a smart meter is installed, the export element is deemed to be 50 per cent of the power generated by the renewable system.
• Your electricity bills are reduced as you use energy produced by the renewable technology.

Domestic technologies qualifying for the scheme include solar PV panels, wind turbines, hydro, anaerobic digestion and micro-combined heat and power. They must be certified by the MCS to be eligible and be fitted by a MCS approved installer.

Where ownership of a property changes, ownership of the generating technology also changes and the FITs payments transfer to the new occupier.

A number of commercial companies offer free solar PV panels under what is known as the 'Rent-a-Roof' scheme. In basic terms, they offer you free solar PV panels and in return the company receives the FIT payments. The Energy Saving Trust website provides information and advice, including some of the questions you should ask if you are considering such an offer.[11]

Warm Home Discount

The Warm Home Discount (WHD) requires energy companies by law to give a discount on gas or electricity bills to some of their most vulnerable customers.[12] It helps around 2 million households in Great Britain annually. All suppliers with more than 250,000 customers must participate, whereas suppliers with fewer customers are able to participate if they choose so.

The government announced in the Spending Review in November 2015 that the Warm Home Discount scheme would be extended to 2020/21 at current levels of £320m per year, rising with inflation, to help households who are at risk of fuel poverty with their energy bills.

Energy companies have been required to provide a discount on electricity bills to a 'core group' of low-income pensioners. It is usually applied between September and March. From 23 July 2016, there is an option to apply the discount to either the gas or electricity bill.

You may qualify for this discount if, on the qualifying date (which is published annually on www.gov.uk):

● your name, or your partner's name, is on your electricity bill; *and*
● you get your electricity from one of the participating energy suppliers; *and*
● you get the guarantee credit of PC (even if you get savings credit as well).

Suppliers are also required to provide the same discount to a 'broader group' of customers such as those on a low income or receiving certain means-tested benefits, although they retain some discretion over who is eligible for the discount. The discount for customers in both the core and broader groups is £140.

If you are eligible as part of the core group you do not initially need to apply for a WHD – you should receive this automatically. Alternatively, you may receive notification from the Department for Work and Pensions (DWP) that you should apply to your supplier (the DWP will provide the appropriate details when it contacts you).

The DWP and participating electricity suppliers are sharing some limited information about their customers. This allows participating suppliers to give the discount automatically to customers who qualify or for the DWP to write to those who may qualify but have not been matched automatically.

Each supplier has slightly different eligibility criteria for its 'broader group' customers, though all have a focus on low-income and vulnerable people.

Contact your electricity supplier to ask if you might be eligible for the discount. WHD is allocated on a first-come, first-served basis.

A list of all suppliers offering WHD is available at www.gov.uk/the-warm-home-discount-scheme/eligibility.

The WHD is one of the obligations which will be devolved to the Scottish government under the terms of the Scotland Act 2016 but details are as yet unknown.

Help with gas connections

The government has set obligations on the eight gas distribution networks (GDNs) in Great Britain to deliver assistance to their consumers. The obligation, known as RIIO-GD1, runs for eight years from 1 April 2013 to 31 March 2021 and is monitored by Ofgem.

As part of this obligation, the Fuel Poor Network Extension Scheme enables customers that are in fuel poverty to switch to natural gas by helping towards the cost of connection to the gas network.

Each of the GDNs covers a separate geographical region and they are owned and managed by the following companies. Contact them to find out what support is offered in your area.

- National Grid Gas plc – East Midlands, West Midlands, North West England and East of England (including North London).
- Northern Gas Networks Limited – North East England (including Yorkshire and Northern Cumbria).
- Wales & West Utilities Limited – Wales and South West England.
- SGN – Scotland and Southern England (including South London).

Green Deal

The Energy Act 2011 made provisions for the introduction of the Green Deal. This UK-wide programme was designed to make energy efficiency improvements affordable by removing the upfront costs. The Green Deal was complemented by the ECO (see p166). In 2015, the UK government stopped funding the Green Deal Finance Company, which was set up to lend money to Green Deal providers.

However, the Green Deal assessment framework is still open and you can still purchase Green Deal assessments. For new applicants, it may still be possible to get Green Deal funding from certain providers. The gov.uk website has a search tool to identify providers near to you still offering Green Deal assessments. A Green Deal assessor will visit your home, talk to you about your property and its energy performance and help you to decide if you need to carry out energy-saving improvements. A Green Deal advice report will be produced, which contains an EPC that rates your home for energy efficiency and also provides an estimate of how much money you could save on your energy bills.

If you already have a loan (ie, a Green Deal finance plan), you are not affected by the latest developments if the improvements to your home have already been made and you currently make repayments via your electricity bill. Your Green Deal provider will continue to be responsible for any warranties or maintenance specified in the contract with them.

If you have a complaint about the Green Deal and also have a Green Deal finance plan, you should contact your Green Deal provider. Its contact details should be on the EPC, which is part of the Green Deal assessment. If your complaint is not resolved within eight weeks then you should contact the Green Deal Ombudsman (email: enquiries@os-energy.org, telephone: 0330 440 1624).

If you do not have a Green Deal finance plan, but experience problems with your Green Deal assessor then contact your provider first. If you are not happy with its response, contact:

- the assessor's certification body; *or*
- Energy Saving Advice Service (telephone: 0300 123 1234); *or*
- Home Energy Scotland (telephone: 0808 808 2282); *or*
- The Green Deal Ombudsman (telephone 0330 440 1624).

The Green Deal did not lend itself particularly well to helping the fuel poor. Many fuel poor households limit their fuel use because of low income. Accordingly, any improvements in household energy efficiency are often taken up in increased comfort levels. Therefore, it may be more difficult for Green Deal assessors to estimate accurately what the potential savings, if any, are for a fuel poor household installing Green Deal measures.

The ECO provides extra help for those most in need such as the vulnerable, those on low incomes and those with homes that are expensive to treat (see p166).

Advice and assistance from suppliers

Providing guidance on energy efficiency for customers is a licence condition for gas and electricity suppliers.[13] Each supplier has to produce a code of practice on using fuel efficiently. It should be published on the supplier's website or you can request a copy by telephone.

Suppliers have trained staff offering advice on ways to save energy and cut your energy bills. In some areas they can also arrange for an adviser to visit your home and make recommendations as to ways of saving energy and money.

You may also be sent booklets on energy efficiency, including details of grants and the supplier's own schemes. Ofgem monitors the energy efficiency measures of the 'big six' energy companies and publishes regular reports.

Many of the suppliers also have funds, foundations and trusts that can provide a range of support and assistance, particularly to vulnerable customers (see p181).

Energy efficiency advice contact details for the big six suppliers

British/Scottish Gas	0800 072 8629
Mon–Fri 8am–5pm	
EDF Energy	0800 096 9966
Mon–Fri 8am–8pm	
Sat 8am–2pm	
E.ON	0345 052 0000
Mon–Fri 8am–8pm	
Sat 8am–6pm	
npower	0800 02 22 20 (landlines)
Mon–Fri 8am–8pm	0330 100 8620 (mobiles)
Sat 8am–6pm	0800 413 016 (minicom/textphone)
ScottishPower	0800 33 22 33
Mon–Fri 8.30am–	
4.45pm	
SSE	0345 076 7638
Mon–Fri 8am–5pm	
Sat 8am–2pm	

2. **Energy efficiency schemes in Wales**

Nest

All households in Wales can access the Nest scheme for advice on fuel tariffs and saving energy and for a benefit entitlement check. In the latter half of 2016, the Welsh government is consulting on the design and delivery of a new scheme to succeed Nest from September 2017.

Those living in the hardest-to-heat private sector properties and who get a means-tested benefit can apply for free home improvements such as a new boiler, central heating system, hot water cylinder insulation or renewable energy technologies – eg, solar panels.

To qualify for Nest home improvements, the property must be:
• your own home, privately rented or shared-ownership – if you rent, you must have the landlord's permission (those who rent from a local authority or a housing association are not eligible);
• energy inefficient (E, F or G rated).

You or someone who lives with you must get one of the following:[14]
• child tax credit and your annual income is below £16,105;
• council tax reduction (discretionary reductions and discounts do not qualify on their own);
• housing benefit;
• income-based jobseeker's allowance;

- income-related employment and support allowance;
- income support;
- pension credit;
- universal credit;
- working tax credit and your income is below £16,105.

If you do not meet the criteria, Nest may be able to refer you to alternative schemes. To apply or to get advice, call 0808 808 2244 (free from a landline or a mobile phone).

Arbed

Arbed is an area-based scheme to retro-fit existing homes with energy efficiency measures and renewable technologies. Arbed means 'to save' in Welsh and the scheme includes measures such as installing new energy efficient boilers, carrying out insulation improvements, installing solar panels and providing energy saving advice.

---

3. **Energy efficiency schemes in Scotland**

Home Energy Scotland

The Scottish government has created a 'one stop shop' – Home Energy Scotland – to communicate its programmes. It is funded by the Scottish government and managed by the Energy Saving Trust via its network of advice centres. The network can help you access grants and offers an energy advice service. All households in Scotland can get a personalised home energy check by calling 0808 808 2282.

Home Energy Scotland Referral Portal

Home Energy Scotland can take referrals from advisers of householders who may be able to benefit from advice and support. Householders can contact Home Energy Scotland on 0808 808 2282 or, with their permission, advisers can refer them through the Home Energy Scotland referral portal.

Advisers may wish to refer a householder who answers yes to any of the following questions.

- Do you find your home hard to heat?
- Do you worry about your fuel bills?
- Would you like advice and support to make your home warmer and reduce your fuel bills?

The online portal is secure and works in real-time. Once advice has been given to the householder, and referrals made to any of the schemes and partner organisations to help with energy saving or income maximisation, the portal reports the outcomes to the adviser.

e,-15 The web address of the Home Energy Scotland portal is https://hespartnerships.est.org.uk.

For more information, or to register as a portal user, contact your local Home Energy Scotland Community Liaison Officer.

Home Energy Efficiency Programmes for Scotland

The Home Energy Efficiency Programmes for Scotland (HEEPS) is the Scottish government initiative to tackle fuel poverty and increase energy efficiency in homes. It is a cluster of programmes currently including:

- area-based schemes, managed by local authorities (see below);
- Warmer Homes Scotland Scheme, managed by Warmworks Scotland (see below);
- a loans scheme, managed by Home Energy Scotland (see p176).

For all HEEPS schemes, the initial access point for information and to check eligibility is Home Energy Scotland on telephone 0808 808 2282 or at www.energysavingtrust.org.uk/scotland/home-energy-scotland. There is a new Scotland's Energy Efficiency Programme (SEEP) expected from 2018.

Area-based schemes

HEEPS area-based schemes provide energy efficiency measures in deprived areas. Local authorities choose which areas are eligible and what measures are available. Whether you qualify depends on your postcode. You may qualify whether you own or rent your home.

Warmer Homes Scotland Scheme

The Warmer Homes Scotland scheme aims to help low-income and vulnerable homeowners and private sector tenants make their homes warmer, more comfortable and energy efficient. Measures available include improvements in central heating systems, insulation improvements, draught proofing and installation of renewable technologies.

You must:

- be a homeowner or the tenant of a private-sector landlord; *and*
- live in the home as your main residence and have lived there for at least six months (unless in receipt of a DS1500 certificate); *and*
- live in home with an energy rating of 64 or lower.

At least one of the following must also apply:[15]

- you or a member of your household or your partner are aged 60 or over and have no central heating and are in receipt of a qualifying benefit;
- you are aged over 75 and in receipt of a qualifying benefit;

- you are pregnant and/or have a child under 16 and are in receipt of a qualifying benefit;
- you have a disability and be in receipt of any level of disability living allowance or personal independent payment;
- you are a carer in receipt of carer's allowance;
- you have been injured or disabled serving in the Armed Forces and are in receipt of Armed Forces Independence Payment/war disablement pension;
- you have an injury or disability from an accident or disease caused by work and are in receipt of industrial injuries disablement benefit. Check www.energysavingtrust.org.uk/scotland/grants-loans/heeps/heeps-warmer-homes-scotland-scheme for the eligibility criteria.

Qualifying benefits

Armed Forces Independence Payment

Attendance allowance

Carer's allowance

Council tax reduction

Disability living allowance

Industrial injuries disablement benefit

Pension credit guarantee credit

Personal independence payment

Universal credit (UC) or a benefit UC is replacing – ie, income-based jobseeker's allowance, child tax credit, working tax credit, employment and support allowance, income support or housing benefit.

War disablement pension

HEEPS loan scheme

Loans are available for owner-occupiers and private sector landlords looking to improve their properties. The loans of up to £15,000 are interest-free and cover a variety of measures such as solid wall insulation, double glazing or a new boiler. The repayment period varies based on the amount borrowed. If you take out a higher value loans, you can pay it back over 10 years.

To make an application, call Home Energy Scotland on 0808 808 2282. Once completed, you can submit your application to the Energy Saving Trust along with your chosen quote. You are strongly advised to seek at least three quotes, particularly if you are considering installing more expensive measures.

You need an assessment to get a Green Deal advice report carried out before work starts because any work must appear on the 'recommendations for improvement' list on the Energy Performance Certificate portion of the report. This can be submitted along with your application form, or can be carried out after your application has been assessed.

Once your application has been received, the Energy Saving Trust assesses your eligibility including undertaking credit and affordability checks. If you are successful, you are offered a loan. Work must not start until you have received a loan offer.

In September 2016, the Scottish government announced a pilot equity loan scheme (part of HEEPS) to run in Glasgow, Argyll and Bute and Perthshire. The scheme provides equity loans to homeowners on low incomes, and small landlords, to help them make essential repairs to leaking roofs and building structures in order to make their homes warmer. This is often essential work that is required before the installation of energy efficiency measures, such as solid wall insulation. Loans of up to £40,000 are available and can be used either as a single equity loan or as part of a package with other Scottish government grants to install more expensive energy efficiency measures. At least 45 per cent of the funding must be for energy efficiency improvements or works that reduce heat loss. 55 per cent or less of the works cost can be used to fund external repairs generally, however cosmetic improvements are not eligible. The terms and conditions are subject to change as the scheme is developed.

Details of repayments are on the Home Energy Scotland website. Applications open in winter 2016 – It is possible to make an expression of interest to Home Energy Scotland on 0808 808 2282. The scheme runs until 31 March 2017, with the possibility of an extension until March 2018.

4. **Help from the local authority**

Home improvement assistance

Local authorities have discretionary powers to improve living conditions in their areas. You may be able to get a grant or discount to help you to improve the energy efficiency of your home. '**Home**' means a property, part of a property, boat or caravan that you live in. Your eligibility depends on what is available from your local authority, and in many areas you must be receiving a means-tested benefit. The local authority must publish details of what is available, who is eligible, how to apply and how to complain.

The Welsh government's 'Houses into Homes' scheme can provide a loan to enable an empty property to be made fit to sell or to let. Contact your local authority for its information pack.

Local authorities in Scotland have similar discretionary powers, under the Scheme of Assistance. The scheme also aims to encourage homeowners to take more responsibility for the condition of their homes, to ensure that private housing in Scotland is kept in a decent state of repair.[16] Contact your local authority to apply for the scheme.

Help from social services

Local authorities have duties to provide services to safeguard and promote the welfare of children in need and promote the upbringing of such children by their families.[17] This could include negotiating with a supplier on your behalf when necessary.

In exceptional circumstances, this can also include providing assistance in cash; a policy not to provide such assistance in any circumstances at all would almost certainly be unlawful, and could be challenged by way of judicial review (see Chapter 14). If such payments are available, you can argue that they can be used to meet all or part of a fuel bill, to buy alternative means of cooking or heating, or to provide other aids for keeping warm, such as blankets.

In Scotland, there are also powers to promote social welfare by 'making available advice, guidance and assistance' to people in need aged 18 or over.[18] This can include giving assistance in kind or, in exceptional circumstances, in cash, where giving assistance would avoid you needing greater assistance from the local authority at a future date.

If you are seeking help from social services in an emergency (eg, because the supplier is threatening disconnection), inform your supplier. Suppliers' codes of practice allow for a delay in disconnection, normally for about two weeks, while a local authority investigates whether it can help, but this delay will only happen if you ensure the supplier knows of the council's involvement.

Social workers may also have good links with and/or be prepared to make referrals to charities for you.

In addition, there is the Scottish Welfare Fund, which provides a safety net for vulnerable people on low incomes through community care grants and crisis grants. This is delivered by all local councils in Scotland on behalf of the Scottish government and is underpinned in law.[19] You may be eligible to receive a grant if you are over 16 and on a low income. To apply for help, you must contact your local council.

See also p200 on local councils' powers in England and Wales to protect occupiers and tenants when an owner or landlord fails to pay fuel or water bills.

5. **Other sources of help**

Energy Saving Trust

The Energy Saving Trust (EST) is an impartial organisation helping people to save energy and reduce carbon emissions. It offers free, impartial advice and information via a range of partnerships. The EST is contracted by the Department for Business, Energy and Industrial Strategy (formerly the Department of Energy and Climate Change) to run the Energy Saving Advice Service (ESAS), a telephone advice service offering impartial energy saving advice in England and Wales. EST also runs Home Energy Scotland, a network of energy centres offering a range of help and support (see p174), under contract to the Scottish government.

The advice service signposts you to organisations that can help with energy saving measures and reduce your fuel bills.

If you live in England or Wales, contact the ESAS on 0300 123 1234. The advice provided is free, but calls are charged at the standard national rates. Home Energy Scotland can be contacted on 0808 808 2282. All calls to this number from landlines or mobiles are free.

Measures to save energy

The EST has produced information on the ways in which you can reduce your fuel bills. Some of these cost-saving measures are listed below.

No-cost measures

Central heating	Turn the thermostat down by 1°C.
Hot water	Set the cylinder temperature to 60°C/140°F (though 63°C–65°C provides protection against health risks, so a lower temperature may not kill all germs during hand washing); use a plug in basin/sink rather than running water.
Curtains	Close curtains at dusk to conserve heat.
Lights and appliances	Turn lights and appliances off when not in use, as standby uses electricity.
	Cool food before putting in the fridge or freezer, and pack empty spaces with crumpled newspaper.
	Fill the kettle with the amount of water you need.
	Wash laundry on low temperatures with full loads.
	Choose the right size pans for cooking, with lids on.

Low-cost measures

Fit energy saving light bulbs.
Insulate hot water tanks and pipes.
Fix dripping taps.
Draught proof exterior doors, letter box and keyholes.
Shower rather than bath.

More detailed information and useful tips on improving your home's energy efficiency can be found on the EST's website.

Citizens Advice

Citizens Advice has been working on the **Energy Best Deal** public awareness campaign with support from Ofgem and major energy companies since 2008. Energy Best Deal has now improved the confidence of over 400,000 domestic energy customers to shop around, reduce their bills and get help if they are falling behind with paying for their fuel.

Practical presentations are delivered to low-income consumers and front-line staff who work with people at risk of fuel poverty. Sessions last about an hour and include topics such as negotiating with suppliers and accessing grants. Information is also provided on where to get further help on benefit entitlements.

Citizens Advice Scotland

Citizens Advice Scotland (CAS) helps deliver the Citizens Advice consumer service, established in 2012 to provide advice on a range of consumer issues. CAS's responsibilities include:

- representing consumers' interests in energy, post and water, working to promote fairer markets and improve customer services in these fields;
- dealing with complex energy cases, or cases received from vulnerable consumers;
- dealing with disconnection or threatened disconnection. These cases may be handled by the **Extra Help Unit (EHU)**, which negotiates with suppliers and offers advice. Most of EHU's cases are referred by the Citizens Advice consumer service, though it may also accept some cases from other sources. EHU can be contacted on 03454 04 05 06;
- hosting an **Ask the Adviser service** (tel: 0344 980 0041), which provides real-time responses to front-line staff to help them deal with various clients' energy issues ranging from industry regulations to metering billing and tariffs.

Priority Services Register

Energy suppliers are required to keep a register of priority service customers who, by virtue of being of pensionable age, disabled or long-term sick or having a hearing or visual impairment, require information or advice on the special services available. See p90 for more information.

National Energy Action

National Energy Action develops and promotes energy efficiency services to tackle the insulation and heating problems of low-income households. It aims to eradicate fuel poverty and campaigns for greater investment in energy efficiency to help those who are poor or vulnerable. See www.nea.org.uk for more information.

Energy Action Scotland

Energy Action Scotland (EAS) campaigns for an end to fuel poverty in Scotland. EAS seeks to develop and promote effective solutions to the problem of cold, damp and expensive-to-heat homes. It provides information and advice to organisations in the fuel poverty field but not to members of the public. See Appendix 1 for contact details.

The Energy Ombudsman

The Energy Ombudsman is an independent body that resolves outstanding energy disputes. It is a free service that deals with complaints about energy companies. See Chapter 14 for more information.

Charities

Some charities, particularly charities for ex-service personnel, offer help to meet fuel bills. It is helpful if an advice agency or social worker can write to the charity to explain your circumstances. The *Charities Digest* (available in reference libraries) lists relevant charities. Another useful book is *A Guide to Grants for Individuals in Need*. Your local reference library also may be able to help locate useful local charities.

Turn2us (www.turn2us.org.uk) is a charitable service which can help you access grants and financial help, and has an online benefits calculator.

However, the demand for charitable payments is high. It is likely that many charities will refuse to help with fuel debts if Fuel Direct or some other budgeting scheme is available. If you are on a means-tested benefit (eg, housing benefit), check that a charitable payment does not affect your benefit.

Trust funds and foundations

Some energy companies have trust funds to help customers who are in debt, or may fund projects which provide support for the fuel poor.

British Gas and Scottish Gas

All customers of British Gas and Scottish Gas with current debt are able to apply to the British Gas Energy Trust Fund. This offers grants to clear arrears of gas/ electricity bills and other essential domestic bills or purchase of essential household items.

See www.britishgasenergytrust.org.uk or phone 01733 421060 for more information.

EDF Energy

All EDF Energy customers with current debt are able to apply to the EDF Energy Trust Fund. The EDF scheme also extends to customers of London Energy, Seeboard Energy and SWEB Energy.

The EDF Energy Trust offers grants to clear gas or electricity debt and to help with other essential household bills and appliance purchases.

Download or complete a form at www.edfenergytrust.org.uk or phone 01733 421060.

E.ON

The E.ON Energy Fund aims to assist E.ON customers who are living in low-income households (a household income of less than £16,190 a year and savings below £8,000). The fund can help pay current or final energy bills arrears, from a previous supplier. It can also help customers buy replacement white goods such as cookers, fridges, freezers and washing machines – and also help to replace and repair gas boilers.

Call 03303 80 10 90 for more information. Calls are charged at standard local rates.

npower

npower's Energy Fund may provide one-off payments (further assistance payments) for household bills and energy arrears.

Apply at www.npowerenergyfund.com or phone 01733 421060.

ScottishPower

ScottishPower's Hardship Fund can help with energy debts. Apply at www.sedhardship.fund or phone 0808 800 0128. In addition, there is the ScottishPower Energy People Trust where not-for-profit organisations can apply for funding to provide support for those in fuel poverty. Priority is given to projects aimed at helping families with young children.

Apply at www.energypeopletrust.com or call 0141 614 4480 or 0141 614 8199.

SSE

No specific scheme is available for individual consumers, but see www.sse.co.uk/help/bills-and-paying for the range of support services available.

Notes

1. **National energy efficiency schemes**
   1 Secretary of State for Energy and Climate Change, *Annual Energy Statement 2014*, October 2014, para 114
   2 DECC, *Annual fuel poverty statistics report*, 30 June 2016
   3 Welsh Government, *The Production of Estimated Levels of Fuel Poverty in Wales: 2012-2016*, 11 July 2016
   4 Welsh Government, *The Production of Estimated Levels of Fuel Poverty in Wales: 2012-2016*, July 2016
   5 Northern Ireland Housing Executive, *Home Energy Conservation Strategy Sixteenth Annual Progress Report*, 2012
   6 NEA fuel poverty statistics, available at www.nea.org.uk/media/fuel-poverty-statistics/

7 Centre for Analysis of Social Exclusion,
   *Getting the measure of fuel poverty: final
   report of the fuel poverty review,* John
   Hills, March 2012
8 Sch 1 Electricity and Gas (Energy
   Company Obligation) Order 2012
   No.3018
9 Product eligibility list at
   www.ofgem.gov.uk/publications-and-
   updates/domestic-renewable-heat-
   incentive-product-eligibility-list
10 www.ofgem.gov.uk/environmental-
   programmes/feed-tariff-fit-scheme/
   tariff-tables
11 www.energysavingtrust.org.uk/
   Generating-energy/Choosing-a-
   renewable-technology/Solar-panels-PV/
   Free-solar-PV-offers
12 The Warm Home Discount Regulations
   2011 No.1033
13 EA1989; condition 31 SLC

2. Energy efficient schemes in Wales
14 Reg 2 The Home Energy Efficiency
   Schemes (Wales) Regulations 2011
   No.656

3. Energy efficiency schemes in Scotland
15 Reg 6 HEAS(S) Regs
16 www.gov.scot/Topics/Statistics/
   Browse/Housing-Regeneration/HSfS/
   SoA

4. Help from the local authority
17 s17 Children Act 1989; s22 Children
   (Scotland) Act 1995
18 s12 Social Work (Scotland) Act 1968
19 Welfare Funds (Scotland) Act 2015

Chapter 13
. . . .

You, your landlord and fuel

This chapter covers:
1. You and your landlord (below)
2. Rent increases for fuel or fuel-related services (p186)
3. Resale of fuel by a landlord (p196)
4. Defective housing and heating systems (see p202)
5. Energy efficiency matters (p212)

1. You and your landlord

Most arrangements for payment of gas or electricity are made directly with the supplier. However, some tenants pay for fuel or fuel-related services (such as heating, cooking, lighting or hot water) indirectly through their landlord – ie, the supplier supplies the fuel to the landlord who resells it to you. Frequently a landlord will:

- provide gas or electricity, pay the bill and recover charges from tenants by sharing out costs on a fixed or variable basis;
- pay the bill and recover charges from tenants by a separate payment system; *or*
- provide heating from a central boiler and recover charges on a fixed or variable basis.

It can be more economical if your landlord provides fuel-related services – eg, a common boiler in a block of flats may be relatively cheap. However, the involvement of your landlord can lead to disputes over the amounts charged or over your position if your landlord fails to pay the bills.

Think carefully before beginning a dispute with your landlord. Always consider the strength of your position. As a tenant, this means considering how secure the tenancy is. This depends on the type of tenancy you have (protected, statutory, assured, assured shorthold, secure or none of these). A full discussion of security of tenure is outside the scope of this book, but it is an important issue because, for example, if you have no security and start a dispute with your landlord, you could end up losing your home. However, in England, the Deregulation Act 2015 has introduced measures to protect private tenants from eviction in retaliation for making requests for repairs or energy efficiency.[1] Protection is given to tenants to make reasonable complaints about the property (including common shared parts) to their landlord without fear that they will be evicted as a result. These rules cover complaints about problems that might cause a potential risk of harm to the health or safety of the tenant, or a family member. Examples of repairs covered include a leak in the property or a problem with the heating, especially in colder weather. You must first make a written complaint to your landlord. If you fail to get an adequate response within 14 days, ask the local authority to pursue the complaint. If you do not have a postal or email address for your landlord and you have made reasonable efforts to contact the landlord, you can still contact the local authority.[2]

In some situations, you have a right of appeal to the courts or a tribunal against the decision of a landlord to charge you for the fuel costs for your home or a failure to permit energy efficiency improvement measures being made to the property. Since 1 July 2013, the First-tier Tribunal (Property Chamber) has the jurisdiction to hear disputes over rent or payments involving a fuel element, with a further right of appeal to the Upper Tribunal on a point of law. In Scotland, a dispute may go to the Private Rented Housing Panel.

If your employer provides accommodation and deducts costs for fuel from your wages, the basic agreement covering such an arrangement is your contract of employment. A written contract of employment should set out the terms and conditions.

Deductions to cover fuel costs from your wages by your employer may be unlawful if your earnings fall below the minimum wage as a result. The Court of Appeal ruled that deductions made for gas and electricity from wages paid to workers at a holiday resort were unlawful where the wages fell below the statutory minimum wage.[3]

Energy performance certificates

Your landlord must provide an Energy Performance Certificate (EPC) providing details of the property's energy use and typical energy costs using an energy efficiency rating from A (most efficient) to G (least efficient). Landlords are required to make EPCs available to prospective tenants at the earliest opportunity.[4] An EPC must be accompanied by recommendations for the improvement of the energy performance of the building to give you an idea of the amount of energy and cost needed to heat the property.[5] If you want to make any of the improvements listed in the EPC, your landlord must consent to certain energy efficiency measures being made (see p212).

An EPC is valid for 10 years and may be supplied either as a hard copy or electronically if you consent.[6] A failure to supply an EPC or a Gas Safety Certificate affects the 'no fault' right of the landlord to recover possession of the property at the county court.

In Scotland, the EPC must be displayed somewhere in the property – eg, in the meter cupboard.

In England, you must also be given a copy of the Department for Communities and Local Government's booklet *'How to rent: the checklist for renting in England'*, which is available at www.gov.uk/government/publications/how-to-rent.[7] Your landlord is not required to supply a further copy of the booklet if a new version is published during your tenancy. This requirement does not apply to registered providers of social housing.

2. Rent increases for fuel or fuel-related services

The circumstances in which your landlord can increase your rent because of increases in charges for fuel or fuel-related services depend partly on whether you have a council or non-council tenancy. If you are a non-council tenant, your rights also vary according to whether you took up the tenancy before or after 15 January 1989 (2 January 1989 in Scotland) – see p194. A landlord's power to increase charges for fuel or fuel-related services can be limited in one of three ways.
- Payments for fuel or fuel-related services are 'service charges', so legislation which affects service charges may be relevant.
- The courts have held that fuel charges are normally part of the rent,[8] so where legislation controls the rent, fuel charges are included.
- A tenancy agreement is a type of contract and may include limits on your landlord's power to increase charges.

If fuel charges are included in your rent, you may be subject to possession proceedings if you fail to pay them in the same way as if you fail to pay your rent. For charges to be recoverable, they must be agreed by both parties at the beginning of the contract or by you both agreeing during the agreement. If you are subject to possession proceedings, always attend any court hearing as the court may proceed in your absence and grant a possession order against you automatically. A landlord proposing to increase the rent or alter the terms of a tenancy agreement must serve a notice under section 6(2) of the Housing Act 1988 in the prescribed form.[9]

Council tenancies

You have a **'council tenancy'** if your landlord is a local authority, unless you have used your 'right to buy' or if, in England and Wales, your tenancy has a fixed term of more than 21 years. Most council tenancies are called **'secure tenancies'**, with changes to rights of succession being made by the Localism Act 2011.[10]

In England and Wales, local authorities are under the general duty to act reasonably in setting levels of service charges or rent for council tenants,[11] and may be subject to judicial review. The Secretary of State has the power to make regulations covering heating charges, but this has not been used.[12]

In Scotland, local authorities are limited to making service charges which they think are 'reasonable in all the circumstances'.[13] There is no definition of 'reasonable' (see p193), but if you think the charges are unreasonable, you can apply for a judicial review (see Chapter 14).

Otherwise, the only protection for council tenants is contractual. If fuel or fuel-related services are provided as part of your tenancy, a failure to provide these is a breach of contract. If there is such a breach, you can go to court to claim damages (ie, compensation) and a court order requiring the local council to obey the terms of the tenancy agreement. Note, however, that legal aid is unlikely to be available for such claims. The terms of your tenancy may be contained in a written statement, in which case any terms relating to fuel or fuel-related services will be clear. However, often not everything is in writing – sometimes there is no written agreement at all. You then have to work out whether your fuel problem is covered by terms implied in your tenancy. An **'implied term'** is one that, although not written down, is considered by the courts to be included automatically in any tenancy.

Every tenancy agreement in England and Wales has an implied term that the landlord will allow a tenant to have 'quiet enjoyment' and that the landlord will not interfere with or interrupt a tenant's ordinary use of the premises. In this case, that would mean not interfering in any way with your use of fuel or fuel-related services. The Scottish equivalent is the tenant's right to full possession of her/his premises, which has the same effect.

In England and Wales, terms are also implied by the Supply of Goods and Services Act 1982 that says that services must be provided with reasonable care and skill, within a reasonable time and at a reasonable charge. What is a reasonable time is treated as a question of fact.[14] Problems with fuel supply or fuel-related services can often come within these terms. In Scotland, similar terms may be implied into the contract by common law.

Local authority heating systems

All local authorities have the power to produce and sell heat, including electricity which is produced by renewable sources.[15] There is no specific protection in relation to heating charges, but the authority must:

- keep a separate account of them;[16] *and*
- when fixing the charges, act in good faith, not for ulterior or unlawful purposes, and within the reasonable limits of a reasonable local council;[17] *and*
- comply with the law on maximum charges for resale of fuel (see p196).

London boroughs have additional powers in respect of the provision of heating by hot water or steam.[18] They may prescribe scales of heating charges that apply, unless there is a specific agreement setting different charges.[19] The charges must be shown separately on rent books, demand notes or receipts, and be differentiated from rent generally.

London boroughs are not allowed to subsidise heating. When providing heat or setting charges, they must not show 'undue preference' or exercise 'undue discrimination'.[20] Some preference or discrimination is inevitable, as not all tenants paying the same charges will be provided with identical heat. To decide if the preference or discrimination is 'undue', consider:

- the cost of providing the heat to you compared with the cost of providing it to other tenants;
- the level and consistency of heat;
- restrictions or terms governing the heat provided – eg, in winter only.

If you can show undue preference or discrimination, you can recover the amount you have been overcharged by taking legal action (see Chapter 14).

If you suspect that the local authority is charging more for heat and power than the actual cost to itself, the actual costs may be obtained by use of the Freedom of Information Act 2000. Under the Act public authorities are bound to make available information within 21 days of a written request, unless the information falls into a number of restricted categories. There is a right of appeal to the Information Commissioner's Office against a refusal to supply information.

Challenging the way heating is provided

If you challenge the legality of the way a heating system is being run or charges for heat, complex legal issues arise. As well as the matters mentioned, a court can consider such matters as whether the local authority charges for:

- assumed heat delivery instead of actual heat delivered, if there is a significant difference;
- heating costs which are significantly higher than those of other heating systems;
- amounts unrelated to heat delivered or assumed to be delivered.

When some heating is provided but it is inadequate, it is difficult to prove that there has been a breach of the tenancy agreement unless there is a specific agreement stating how much heating is to be provided and at what times of the year. If nothing is specifically agreed or set out in the tenancy agreement, there is probably an implied term that 'reasonable heat' should be provided, but this is extremely vague. If there is a dispute, keep a detailed diary of when the heating was sufficient, when it was inadequate or off altogether, and even when there was too much.

A failure to consult adequately on local authority changes to district heating schemes is potentially open to judicial review.[21] In respect to specific groups of dwellings, the local authority or landlord may apply to the First-tier Tribunal (Property Chamber) for a dispensation from the duty to consult – eg, when replacing boilers on an estate. If a landlord seeks a dispensation, you may challenge it by showing that you will be prejudiced, either financially or otherwise.[22] Once you – and other leaseholders – have shown a credible case for prejudice, it is for the landlord to rebut it, and the tribunal will regard the leaseholders' arguments sympathetically. The duty to consult is an important one and good practice would dictate that it occurs even in emergency situations. Failures to consult and delays in doing so will count against a landlord.[23]

Pressure by tenants' groups
It may be more effective for tenants' associations to put pressure on a local authority to change the way it manages the heating system or the charges for it. In challenging high heating charges, tenants' groups can look at:
– copies of local council committee reports on heating systems and charging policies;
– a comparison of income from, and expenditure on, individual estate systems and across a local council area;
– expenditure charged to the heating account: does it include all fuel expenditure, maintenance, insurance, caretakers' wages, interest on the cost of the system; is this consistent with other public landlords?;
– district heating systems: the number of dwellings supplied, the costs and type of fuel used;
– level of service: heating and hot water, hours per day, winter and summer, temperature standards assumed and achieved;
– method of calculation of charges: pooling of costs, property by property, flat charge, charges related to size and number of bedrooms;
– energy efficiency of dwellings: insulation quality, double-glazing;
– arranging a temperature survey to find out what heat is being delivered. Temperatures in all rooms at different times of the day can be measured simultaneously in a number of dwellings.
Requests made under the Freedom of Information Act 2000 may assist in obtaining relevant information from local authorities (see p188).

Heating standards

In England and Wales, your home has to have adequate provision of heating if it is to be regarded as 'fit for human habitation'. According to the government's guidance, **'adequate provision'** means heating which provides a temperature of 18°C in the main living room and 16°C in all other rooms when the outside air temperature is -1°C.[24] However, a local authority may provide heating to another standard which it has set for itself. These are examples of standards in use:

- Chartered Institute of Building Surveyors: from the mid-1970s, recommended 21°C in living rooms, 18°C in kitchens and 16°C in hallways and bedrooms;
- British Standards Institution (Code of Practice BS5449): 21°C in living rooms and dining rooms, 22°C in bathrooms, 18°C in bedrooms, kitchens and toilets, and 16°C in hallways.

In Scotland, a property is considered uninhabitable if is deemed 'below tolerable standard', which may include lacking in satisfactory provision for heating. Local authorities may take action regarding properties that fall below this standard under powers contained in Housing (Scotland) Act 2006.

Some landlords use their own standards. Ask your local authority what standards it uses, as these are probably used in setting the charges.

Non-council tenancies

If your landlord is not a local authority, legislation on variable service charges and on rent control applies. The legislation on variable service charges does not apply in Scotland (see p193), but there are some court cases which give rights to tenants in this area. The provisions for rent control are different for all tenancies granted before 15 January 1989 compared with most of those granted after 15 January 1989 (2 January 1989 in Scotland).

Variable service charges in England and Wales

In England and Wales, variable service charges are covered by the Landlord and Tenant Act 1985.[25] If your landlord used to be a council but it sold the property to a private landlord, you have similar rights under the Housing Act 1985.[26]

Variable service charge

This is an amount payable by a tenant as part of or in addition to rent – directly or indirectly for services – the whole or part of which varies according to the landlord's costs or estimated costs.[27]

This is a broad definition and includes payments for fuel, whether made directly to the landlord or indirectly through a landlord's meter. These provisions apply to all tenants unless you are:

- a tenant of a local council or any other public authority, unless your lease is for over 21 years or was granted under the 'right to buy' legislation;[28] *or*
- a tenant whose rent has been registered with a service charge as a fixed sum.[29]

Your landlord can recover the costs of the services s/he provides (eg, as heat, light or cooking facilities) only if the service is of a 'reasonable' standard and the costs are 'reasonably' incurred.[30] There is no one definition of 'unreasonable', but it includes something which can be proved to be excessive. What is reasonable is a question of fact and degree.[31]

If the charges are based on an estimate in advance, the estimate must be reasonable and, after the costs have actually been incurred, the charges must be adjusted by repayment, reduction of future charges or additional charges, whichever is appropriate. If they are paid in arrears, most charges cannot relate to periods of more than 18 months before.[32] Similarly, charges may not be reasonable to impose where major works are undertaken and charged to an individual tenant or leaseholder who may only have a short period left in occupation of the property.

You are not liable for any costs included for any service charge which was incurred more than 18 months before a demand for payment of the service charge is served. Only if you are served with a notice in writing during the 18 months that the costs have been incurred, and you are required to meet them, does a right to recover the charges arise.[33] In the context of energy bills this is likely to be the point at which the charges are identified as becoming payable, not necessarily when the services were provided or used. For example, in *OM Property Management Ltd v Burr*, the management of an estate received and paid gas bills from the wrong energy company. The mistake was found some years later and a higher bill had to be paid to the actual gas supplier, with costs passed on in the service charges. The Court of Appeal ruled that, although the intention of Parliament was that tenants should be protected against stale claims, the argument that costs recoverable as a service charge are incurred when services are used was rejected. A liability did not become a recoverable cost until it was established, either by being met or paid or possibly by being set down in an invoice or certificate under a building contract. The Court as also noted that estimated costs could be legitimately included in a service charge.[34]

Any obligation to pay a service charge must be founded on the terms that the parties have expressly agreed and recorded in writing. Agreements must be read and understood in their proper context, as they would be understood by any objective reader of the lease who was aware of the circumstances when the lease was entered into.[35] The cost of providing services may include the cost of administration and overheads but the lease must be worded appropriately to allow this.[36] The fact that a management company undertakes administrative work in paying an electricity bill for common parts does not mean that a management fee is recoverable unless management fees are expressly stated as recoverable in the lease.

In the case of service charges applied for caravan pitches, electricity charges are not recoverable unless expressly mentioned in the agreement for the pitch.[37]

If you think service charges should not be payable, apply to the First-tier Tribunal (Property Chamber) to rule on these questions.[38] There is no prescribed form of application. A letter may suffice but it should be marked 'Notice of Application' and must include:

- your name and address and that of any representative;
- an address where documents for you may be sent or delivered;
- the name and address of the landlord and any other respondent (eg, a property management company) and other interested persons;
- the address of the property concerned;
- details of what you are seeking and reasons for making the application;
- a statement that you believe that the facts given in the application are true; *and*
- your signature and the date.[39]

Seek advice before commencing an application. A fee of £100 is payable for making an application, but this can be remitted if you have financial difficulty in paying it.

It may be possible to negotiate with the landlord after making an application. A landlord may agree to reduce the charges before going as far as the tribunal hearing.

Access to information

The right to challenge unreasonable charges would be almost useless without access to supporting information about how the charges are made up. You have the right to require your landlord to provide information; your request must be in writing. Your landlord must provide a written summary of costs incurred over 12-month periods, and must comply within six months of your request.[40]

If the service charges are payable by tenants of more than four dwellings together, the summary of costs must be provided by a qualified accountant.[41] This is aimed mostly at tenants such as those in mansion blocks, but it also applies if you live in a house in multiple occupation.[42]

Within six months of receiving the summary of costs, you can require your landlord to allow you to inspect accounts and receipts. You can also make copies of any documents at a reasonable charge.[43] This is particularly useful if you suspect you are being overcharged.

It may be more effective to exercise these rights through a tenants' association. If members' tenancies require them to contribute to the same costs, a tenants' association can apply to the landlord to become a 'recognised tenants' association'.[44] If the landlord does not agree to this, the association can apply to the local rent assessment committee for a certificate requiring the landlord to recognise it. It can then exercise the rights to information on behalf of its members.

County court proceedings

The county court also has the power to make declarations on the same points[45] under the Landlord and Tenant Act 1985, but you are normally expected to use the First-tier Tribunal. The county court has the power to transfer proceedings to the First-tier Tribunal where a question within the jurisdiction of the tribunal arises.[46] Both the court and the tribunal have powers to deal with litigation costs incurred in proceedings or subsequent to the transfer but only the county court can reduce or extinguish costs incurred in the county court.[47] This is a matter that the court should take into consideration when deciding whether or not to ask for or agree a transfer.

Variable service charges in Scotland

The legislation mentioned above for variable service charges does not apply in Scotland. To find out if there is any limit on your landlord's discretion to increase charges for fuel or fuel-related services, look at your written tenancy agreement if you have one. If there is a term which covers how service charges can be increased, that applies.

If there is no such term or it is unclear, then the courts may be prepared to introduce an 'implied term' (see p187) into the tenancy agreement. In one case,[48] the court introduced an implied term that any service charge had to be 'fair and reasonable'. The court also decided that a surveyor's certificate claiming that the charges were reasonable was not valid because it was the landlord himself who signed the certification in his role as the surveyor.

Rent control

If it is not clear what kind of tenancy you have, refer to any standard text on the law of landlord and tenant.

Rent control is relevant to payments made to a landlord for fuel and fuel-related services because such payments are normally part of the rent; therefore, the payments can be increased only if the rent can be increased. If you get housing benefit (HB) and have a shortfall paying your rent, you may apply for a discretionary housing payment (DHP – see p157). This does not cover the costs of heating, which are excluded from HB, but relates to your overall level of rent. Costs of living, including heating, can be a relevant factor the local authority can consider when deciding whether to pay you a DHP.

The Rent Act 1977 and the Rent (Scotland) Act 1984 used to provide a comprehensive system of rent control. However, they do not apply to most tenancies that started after 15 January 1989 in England and Wales, or 2 January 1989 in Scotland, as these are covered by the Housing Act 1988 or the Housing (Scotland) Act 1988. The system of rent control under these later Acts is so loose that it is virtually useless as a tool for limiting rises in charges for fuel or fuel-related services and is not, therefore, covered in this book. However, from 2017 a degree of rent control is being reintroduced in Scotland following the Private Housing (Tenancies) (Scotland) Act 2016 with a tenant having the right to refer any increase to a Rent Officer.[49]

Tenancies granted before 15 January 1989 (England and Wales) or 2 January 1989 (Scotland)

There are different rent control systems for:

- tenancies which are regulated – ie, protected or statutory; *and*
- tenancies which are restricted/Part VII contracts. It is outside the scope of this book to discuss the different types of tenancy, but a restricted contract can be said to include most tenancies where the landlord is resident;
- tenancies where some form of board (eg, breakfast) is provided – these are excluded altogether from this book.⁵⁰

Tenancies starting after 15 January 1989 (England and Wales) or 2 January 1989 (Scotland) are also regulated if they were granted:

- to an existing protected or statutory tenant (ie, regulated under the Rent Act 1977 or the Rent (Scotland) Act 1984); *or*
- as a result of a possession order made against an existing statutory tenant in 'suitable alternative accommodation' proceedings; *or*
- in accordance with an agreement made before 15 January 1989 (England and Wales) or 2 January 1989 (Scotland).

Regulated tenancies

If you are still a regulated tenant under the Rent Act 1977, your position depends on whether or not there is a registered fair rent (see below). Once a tenancy has been granted, it is contractual until the agreed period ends (usually by a notice to quit). When this happens, a statutory tenancy arises so long as you continue to occupy the premises as your residence.

Unregistered rent

- **Contractual.** If, at the time you were granted a tenancy, no rent had been registered, then there was no restriction on what rent might be agreed. Once agreed, the rent – including any charges for fuel or fuel-related services – can only be raised if:
  - there is a rent review clause (unusual in a residential tenancy); *or*
  - it is done on an application to register the rent; *or*
  - you enter with your landlord into a 'rent agreement with a tenant having security of tenure'.⁵¹ Essentially this is a new tenancy agreement, so you are effectively starting from scratch.
- **Statutory.** When the statutory phase begins, the last contractual rent applies, but may be changed in accordance with changes in the provision or costs of services (including payments for fuel or fuel-related services) or furniture.⁵² There is no set form for giving notice of any change. If you do not agree in writing to any such proposed changes, you or your landlord can apply to the county court (the sheriff court in Scotland) to determine the change.⁵³ The court can take into account past changes when deciding on any increases.

Rather than bothering with applications to the court, if you have a serious disagreement with your landlord, it is more sensible to apply for a fair rent to be registered (see below). Alternatively, you and your landlord can enter into a 'rent agreement with a tenant having security of tenure' and the tenancy reverts to being contractual.

Registered rent

You, your landlord or both of you together can apply for a fair rent to be registered.[54] The registered rent must include any sums payable for services, including payments for fuel or fuel-related services.[55] It does not matter whether you make such payments to your landlord at the same time as payments for rent, at different times, or under separate agreements.

The rent officer must include service payments in the total figure for the registered rent but must state them as a separate figure.[56] Rent officers should also separately note costs in rent which are ineligible for local housing allowance and HB – these include gas and electricity costs, heating, hot water, lighting and cooking.[57]

The rent that is registered for fuel and fuel-related services should reflect not the cost to your landlord, but their value to you. For example, if your central heating system works poorly, the payments to your landlord can be reduced – by 50 per cent in one reported case.[58] You can argue that if your heating system runs efficiently, but is expensive because of the type of fuel used (eg, underfloor heating), your rent should be reduced to the level of a system which provides the same heat at a cheaper rate. Similarly, if the premises are poorly insulated or use energy inefficiently, this could be reflected in your rent.

The amount specified for services in the registered rent is a fixed sum unless your tenancy has a variation clause relating to payments for services, and the mechanism for variation is reasonable (again, there is no definition of 'reasonable').[59] If there is such a clause, the amount entered in the rent register can be a sum which varies in accordance with the terms of the clause. You are then protected by the rules governing variable service charges (see p190).

For more information and to access the electronic rent register, see the Valuation Office Agency website (www.voa.gov.uk).

Sheltered accommodation and fuel charges for communal areas

HB is not normally available towards the heating cost element in service charges but an exception is made for communal areas in sheltered accommodation – ie, common access areas such as hallways and passageways and rooms that are in common use by occupiers of the sheltered accommodation.[60]

Restricted contracts

The level of rent is fixed by the terms of your tenancy contract, which may be oral or written. You or your landlord may refer the contract to an independent tribunal. Since 1 July 2013, this has been to the First-tier Tribunal (Property Chamber) with a further right of appeal to the Upper Tribunal on points of law.[61] The First-tier Tribunal has the power to increase or reduce the rent to a level that it considers reasonable in all the circumstances.[62] There is no definition of 'reasonable', but a 'reasonable rent' is not the same as a 'fair rent' and is normally higher. These are known as 'Part VII contracts' in Scotland and from 2017 will be governed by the Private Housing (Tenancies) (Scotland) Act 2016.

The rent set by the tribunal includes an assessment of any payments for services including fuel or fuel-related services. Once the rent is set, this applies for two years unless you and your landlord agree to a new application, or apply on the basis of a change in circumstances.[63] A change in the services provided is a change in circumstances; a simple increase in gas or electricity prices is probably not a sufficient change unless it was entirely unforeseen when the rent was set.

3. **Resale of fuel by a landlord**

There are specific controls on the maximum charge for gas or electricity supplied to landlords and resold to tenants. The basis on which your landlord sells you fuel is a term of your tenancy agreement – in practice, this is not normally set out in writing. However, it is subject to an upper limit – a landlord reselling fuel cannot recover more than the maximum charge.

Maximum permitted charges

Ofgem has the power to fix maximum charges for the resale of gas and electricity, and must publish details of any charges fixed.[64] This power was exercised to give directions in 2003. Charges have two elements: a charge for each unit of electricity or therm of gas consumed and a 'daily availability charge' to cover the standing charge.

Ofgem has the means to set the method for fixing the maximum price that may be charged by a landlord.[65] The maximum amount should be no more than the price the landlord pays for the supply, although the landlord may offer more than one rate if you have a choice to take fuel or not.[66] Ofgem is able to set prices in a way that will allow any benefits obtained by the purchase of cheaper electricity to pass through to the consumer, in the same way that currently happens with gas. Your landlord can be required to provide certain information to you. In addition, it is possible for you to recover interest on any charges levied in excess of the maximum retail price.

If your landlord overcharges for gas or electricity, you can recover the excess through legal action through the civil courts in a claim for the amount overcharged and interest. If you are overcharged, make reference to Ofgem's direction setting out the latest maximum resale price and guidance.[67] Interest (at a rate of 1 per cent less than the base rate of Barclays Bank plc) is payable on any overcharge for gas supply.[68]

If your landlord undercharges, you may have to make an additional payment, but this depends on your tenancy agreement. There is no implied term (see p187) that you should pay the maximum charge. The provisions on reselling do not apply to consumption charges for powering electric vehicles.[69]

Approval of meters

Electricity meters cannot be used unless the pattern and the method of installation are as approved by regulations.[70] The meter must be tested and approved by a meter examiner appointed by Ofgem. Ofgem has the power to prosecute and fine your landlord for failure to comply with these provisions.

Gas meters must be of a pattern approved by the Gas and Electricity Markets Authority and stamped by, or on behalf of, a meter examiner appointed by the Authority.[71] A supply of gas through an unstamped meter is an offence subject to a fine. The Gas Act does not state who would prosecute, but presumably it would be Ofgem. There is a power to make regulations for re-examining meters already stamped and for their periodic overhaul, but none have yet been made.

Obtaining a meter directly from a supplier

If you encounter continual problems with your landlord's approach to reselling fuel, you could get your own supply directly from a supplier. Both gas and electricity suppliers are under an obligation to provide a supply, with your own meter, if requested to do so by a consumer, although you may have to pay connection charges. If you are doing this because of persistent breaches of the tenancy agreement by your landlord, you may be able to recover the charges from the landlord as compensation for the breaches.

If you have a meter installed, this would be a tenant's **'improvement'** – ie, an 'alteration connected with the provision of services to a dwelling house'. A tenant of a secure or regulated tenancy is not allowed to make any improvement without the consent of the landlord.[72] Landlords cannot withhold their consent unreasonably. If it is unreasonably withheld, it is treated as given. If suppliers are reluctant to co-operate with you, remind them that you have these rights.

The landlord fails to pay bills

If you pay for fuel with your rent, you can be disconnected if your landlord does not pay the bill. The supplier's codes of practice should lay down a period during which disconnection action will not proceed in such circumstances. There are a number of legal remedies to deal with conflicts in this area (see Chapter 14).

The landlord is disconnected

Gas can only be disconnected at the premises for which there are arrears.[73] However, electricity can be disconnected at any premises which your landlord occupies and for which s/he is registered as the consumer (eg, at the landlord's home or workplace) for failure to pay a bill incurred at other premises.[74] Your electricity supply may, therefore, be at risk of disconnection if your landlord has arrears elsewhere. This provision is more difficult to use if the two premises in question are supplied by different companies. Remember that the supplier may not be aware that your landlord is not the occupier unless you, as tenant, provide this information. Inform the supplier of the situation. Always press for disconnection of your landlord rather than you, if disconnection cannot be avoided.

Breach of quiet enjoyment

If you pay for fuel with rent, it is a term of your tenancy (implied, if not written down) that your landlord maintains the supply. If your landlord fails to pay a bill and the supply is threatened or cut off, you could seek a court order to restore the supply and for damages for loss and suffering. Where action or inaction (such as failing to pay a bill) by a landlord results in disconnection, the landlord is in breach of an implied covenant to ensure a supply of gas and electricity and for breach of the implied covenant for 'quiet enjoyment'[75] (in Scotland, for having been deprived of full possession).

The covenant of quiet enjoyment protects you against both wrongful acts by a landlord and also lawful acts of other persons claiming under the landlord, by way of entry, eviction or disruption of your peaceful enjoyment of the land. Thus, interventions by fuel suppliers which have been caused by the landlord's wrongful act or omission may count as breaches of quiet enjoyment. You can sue the landlord for breach of the implied term of contract that s/he would supply the fuel through the meters so long as the tenancy continued, and also for breach of the covenant of quiet enjoyment. This covenant is 'not confined to direct physical interference by the landlord but extends to any conduct of the landlord or his agents which interferes with the tenant's freedom of action in exercising his rights as tenant'. For example, if a fuel supplier takes lawful action to disconnect because of a bill which a landlord has failed to pay, a claim for breach of quiet enjoyment is sustainable against the landlord for having allowed the situation to arise. A claim for distress arising from disconnection may be included.[76] Such a claim may cover physical inconvenience and discomfort caused by the breach and mental distress directly related to that inconvenience and discomfort[77] and, if the disconnection is deliberate, a claim for harassment may also be added. However, a claim for distress will not extend to periods where you are not in actual occupation or for when third parties (eg, family members or friends) may be in occupation instead of you.[78]

Protection of supply where a tenant is insolvent

Where you are unable to pay for fuel supplied by a landlord because of bankruptcy or insolvency, the landlord cannot disconnect the supply to compel you to pay outstanding charges for fuel.[79] The supplier is not entitled to make it a condition of the continued supply that any outstanding charges incurred before the insolvency are paid. See CPAG's *Debt Advice Handbook* for more about dealing with debt.

Protection of supply where a landlord is insolvent

If your landlord becomes bankrupt, the official receiver (or an insolvency practitioner) automatically becomes landlord of the property upon his/her appointment as trustee of your landlord's estate. You must be notified of this in writing. Questions about the supply should be directed to the trustee.[80] Where you have an agreement on something such as the heating system, the official receiver should obtain all documents relating to that agreement and consider continuing the agreement if the cost of doing so is not prohibitive.

Breach of trust

If two or more tenants contribute to the same costs by paying a variable service charge, the sums paid to the landlord are held 'on trust' by her/him. This imposes strict obligations on the landlord as 'trustee'. Failure to pay fuel bills with this money is a 'breach of trust'.

Harassment

Your landlord is committing a criminal offence if s/he harasses you in order to make you give up your tenancy or prevent you from exercising your rights. 'Harassment' means action likely to interfere with your peace or comfort, including the withdrawal of services such as the supply of gas or electricity.[81] The offence may be committed by the landlord of a residential occupier (which is wider in meaning than a tenant) or by an agent of the landlord.

A person convicted of an offence of harassment in the magistrates' court may be jailed for up to six months or fined up to £5,000; if convicted in the Crown Court, the prison sentence may be up to two years and the fine at whatever level the court sees fit. Where you are the victim of unlawful harassment, you can also sue for damages, which can be very large if you have to give up your home.[82] You may be able to obtain advice and assistance from the tenancy relations section of the local authority.

Proceedings may be taken in the county court or High Court. An injunction may be obtained from the court, ordering a landlord to restore fuel supply (see p236). In emergency cases, the injunction may be obtained outside normal court hours by telephoning the court. Damages are also available for breach of contract and for harm caused by acts of harassment, of which the cutting off or disruption of fuel supplies may be just a part. This area of law is governed by the law of tort which covers civil harms, wrongs and injuries. Four different types of damages may be available in a case of harassment or unlawful eviction, depending on the facts of the case. Potential claims may include:

- special damages, representing financial loss that can be identified – eg, cost of alternative accommodation;
- general damages, to put you back in the position you would have been in if the harassment or eviction had never happened. These include damages for pain, distress and nuisance;
- aggravated damages, which are awarded for especially severe harm and demonstrate the outrage and indignation of the court at the conduct of the landlord;
- exemplary damages, awarded where a landlord has acted with a deliberate disregard for your rights and her/his behaviour is calculated to make a profit. Exemplary damages are awarded where it is necessary to 'teach a wrongdoer that tort does not pay'.[83] To sustain an action for exemplary damages, other torts such as trespass, assault and nuisance would have to be shown in addition to action for breach of quiet enjoyment.[84]

Your landlord has a defence if s/he can show that s/he had reasonable grounds for interfering with your peace or comfort, or for withdrawing services – eg, the gas was turned off because of an emergency such as a nearby fire.

Local authorities are often prepared to prosecute landlords for harassment. You can ask the tenancy relations officer to intervene.

Transferring the account

If your landlord consistently fails to pay bills, the simplest solution may be for you to open an account and get the supply in your own name (see Chapter 3).

Rewiring work might be necessary if a meter is moved or a new one installed. For example, in houses in multiple occupation, considerable work is needed to replace one main meter with separate meters for each tenant – in most cases, you would only have to pay for the costs of work to the premises you yourself occupy.

If a supply is being transferred because of a breach of the terms of the tenancy, you can claim the costs of the work as damages in a court action. Otherwise this work is an 'improvement' (see p205).

Local authorities' powers in England and Wales

Local authorities have powers to help tenants whose landlords are endangering their gas or electricity supply through non-payment of charges. If you are seeking the help of the local authority in an emergency (eg, the supplier is threatening disconnection), tell the supplier – its code of practice may allow for a delay in disconnecting while the council investigates whether it can help, but the delay will only come into effect if the supplier knows that the council is involved.

Note: although the following section on local authorities' powers has been written for landlords and tenants, it applies where any 'occupier' has been affected by the failure of an 'owner' to pay a bill. 'Owner' has a wide definition,[85] and might cover the position where one of a number of flat-sharers is both tenant and the person responsible for paying the fuel bills.

Outside London

Local authorities outside London have the power to protect occupiers if the supply is threatened or cut off as a result of an owner's failure to pay fuel charges.[86] Once a request is made in writing, the council can make arrangements with the supplier to reconnect the supply; such arrangements can include payment of arrears and disconnection or reconnection charges.

Having arranged reconnection, the council can recover expenses plus interest from the person who should have paid in the first place. If you have an arrangement where the owner, your landlord, pays the fuel bills, the local authority can also serve notice on you to pay your rent directly to it to set off against its expenses.

Within London

London boroughs have powers to protect you where an owner, usually a landlord, fails to pay a bill.[87] They can make arrangements with suppliers, including paying the expenses of reconnecting the fuel supply. After reconnection, they have a duty, as long as they think it is necessary, to pay the supplier's charges for future consumption.

However, the boroughs have no power to pay arrears – ie, for past consumption. This can be a stumbling block, as suppliers are under no obligation to reconnect the supply while money is owed. The supplier cannot chase you for the arrears because your landlord is the customer, not you. While such arrears are outstanding, a supplier may be reluctant to reconnect a supply.

You can get round this by getting the borough to recover the arrears. The borough has the power to take proceedings to recover money owing at the time fuel was reconnected,[88] and these proceedings can be taken against either you or the defaulting owner. If you pay your rent to the borough under these provisions, you are treated as meeting your obligation to pay rent to the owner. You cannot be required to pay more than the rent that you would otherwise be paying to your landlord.

Suppliers should be keen on this kind of arrangement and prepared to reconnect, as it means the borough does the supplier's debt collecting and, provided you pay your rent to the borough, payment is guaranteed. Boroughs can protect themselves by 'registering a charge'[89] on the affected property to recover their expenses (including administrative costs). Local authorities usually appoint a particular officer – such as a tenancy relations officer – to deal with these matters. If they are reluctant to become involved, you can point out that they can put a charge on the property to cover their expenses and protect themselves. It is unlawful to have a blanket policy not to exercise these powers; they must consider each case individually. If you are told, 'We don't do that', you can consider judicial review (see Chapter 14).

Local authorities' powers in Scotland

There are no powers in Scotland equivalent to those in England and Wales. However, local authorities throughout Great Britain do have powers to make 'control orders' in extreme cases.[90] This means that the council can take over a house in multiple occupation from a landlord and collect the rent in order to pay for any necessary repairs and to pay bills such as fuel bills. Unfortunately, it is unlikely that a control order would be made on the basis of unpaid fuel bills alone; these powers are more typically used if there is substantial disrepair.

The scope of the powers given to local authorities in Scotland under the Housing (Scotland) Act 2006 has yet to be tested with respect to non-payment of fuel charges. A new system of tenancies for privately rented dwellings comes into operation in 2017 following the enactment in April 2016 of the Private Housing (Tenancies) (Scotland) Act 2016.

Social landlords

'**Social landlords**' are local authorities, housing associations and housing co-operatives. They provide housing mostly for people on lower incomes. Some have set up schemes for their tenants.

The former regulator Ofgas produced a useful guide for social landlords on all aspects of the competitive market and Ofgem continues to apply these principles.[91] This is also relevant for tenants' associations which might want to work with their landlords to promote or introduce some of its ideas. The guide mentioned three categories of action which social landlords are taking.

- **Marketing alliances.** The landlord finds a supplier which it thinks provides a good deal for tenants and then works with the supplier to promote the deal to tenants. In return, the supplier may offer discounts or other benefits.
- **Energy service companies.** These help to fund energy efficiency improvements which should pay for themselves in lower fuel bills.
- **The landlord as an energy supplier.** Social landlords can set up a company which applies for a licence to become a gas or electricity supplier itself. Rather than make a profit, supply to tenants can be at cost price.

4. **Defective housing and heating systems**

A full discussion of legal remedies for defective housing is outside the scope of this book. However, defective heating systems and appliances, structural disrepair, use of poor materials, and inadequate insulation and draught proofing can all contribute to high heating bills. Tackling these problems can be expensive and is

rarely a tenant's responsibility. Therefore, this section looks briefly at the legal remedies available to a tenant, dividing them into:
- repairing obligations;
- negligence;
- premises prejudicial to health;
- other local authority powers.

There are also regulations covering the maintenance of gas appliances by landlords (see p211).

There are different forms of action which can be taken against a landlord, but the purpose is always to get work carried out and/or to get compensation. Good records are important evidence and can make a big difference to the level of any compensation. Keep proper records of what is in disrepair and, for instance:
- when the problems started;
- when your landlord was first told of the disrepair;
- all other occasions on which your landlord has been told about the disrepair;
- what has been done, if anything, to put things right.

If you incur extra expenses (eg, to keep warm, eating out or for replacement heaters), make notes and keep receipts. If heating bills are higher than normal, also keep these. The sums can be recoverable.

Protection under the Deregulation Act 2015

The Deregulation Act 2015 protects tenants in England against unfair eviction.[92] Where your landlord fails to address a genuine complaint you have made about the property's condition, and the complaint has been verified by a local authority inspection, your landlord cannot evict you using the 'no fault' eviction procedure (known a section 21 eviction[93]). A 'no fault' eviction is one where the tenant does not have to have done anything wrong. The landlord is also required to ensure that the repairs are completed and you may also have a counter-claim for damages arising from the failure to undertake repairs.

This is being introduced:
- from 1 October 2015, for new or renewing assured shorthold tenancies;
- from 1 October 2018, for all assured shorthold tenancies.

Repairing obligations

Your landlord's repairing obligations may be set out in a written tenancy agreement; for the vast majority of private tenancies entered into since 1989 in England and Wales, this is an 'assured shorthold' tenancy or, more rarely, an assured tenancy. Tenants of councils and housing associations in Scotland have a right to a formal written lease.[94] Whether or not you have a written agreement and whatever is stated in any such written agreement, there is legislation which puts a wide range of obligations on landlords.[95]

Repairing obligations can be enforced through the county court and apply against both public and private sector landlords. For claims of up to £10,000, the arbitration or small claims procedure is used (see p231). As a well as compensation, remedies by way of specific performance or an injunction are available to compel a landlord to fulfil obligations or take steps to remedy harm.[96] Under the rules of court, you are expected to take certain steps before commencing a claim. A 'pre-action disrepair protocol' applies, setting out procedures for both parties to follow in a disrepair claim, encouraging exchange of information and settlement without recourse to litigation. Details and guidance notes are available on the Ministry of Justice website (www.justice.gov.uk).[97]

Scotland

The landlord's obligation is to make sure that any property s/he rents out is in a 'tenantable and habitable condition',[98] or 'reasonably fit for human habitation'.[99] These two phrases almost certainly mean the same thing – the property must be safe, free from damp and generally in a suitable condition for you and your family to live in. Local authorities may act where properties are 'below tolerable standard'. Private landlords in Scotland have a duty to ensure that rented accommodation meets a basic standard of repair called the 'Repairing Standard' under the Housing (Scotland) Act 2006.[100] This covers the legal and contractual obligations of private landlords to ensure that a property meets a minimum physical standard.

Your landlord must carry out a pre-tenancy check of the property to identify work required to meet the Repairing Standard and notify you of any such work. Your landlord has a legal obligation to provide you with written information about the effect of the Repairing Standard provisions on the tenancy. A home meets the Standard if:

- it is wind and water tight and in all other respects reasonably fit for human habitation;
- its structure and exterior (including external pipes) are in a reasonable state of repair and in proper working order;
- its installations for the supply of water, gas and electricity and for sanitation, space heating and heating water are in a reasonable state of repair and in proper working order;
- any fixtures, fittings and appliances provided by the landlord under the tenancy are in a reasonable state of repair and in proper working order.

It is the landlord's duty to repair and maintain the property from the tenancy start date and throughout the tenancy. This includes making good any damage caused by doing this work. On becoming aware of a defect, the landlord must complete the work within a reasonable time.

The repairing obligation applies to all tenancies except Scottish secure tenancies and short Scottish secure tenancies with various social landlords and certain agricultural tenancies,[101] and has been updated from December 2015. The landlord should inspect the property and bring it up to standard before any tenancy starts. If this is not done, you can sue for damages and/or an order of 'specific implement' to force the landlord to carry out any necessary works. The landlord's duty may include carrying out works to improve the property, rather than merely repairing it, if that is necessary to comply with her/his duty. However, it is much more difficult in Scotland to get an order ('specific implement') which enforces that duty.

If problems arise after the tenancy starts, your landlord is only obliged to deal with them if s/he knows, or should know, about them. This means you should report any problems as soon as they arise, preferably in writing. If you believe your home falls short of the Standard, contact your landlord. To be able to take follow up action you must submit your request and claim in writing and get proof of posting.

If you cannot agree with your landlord about whether or not the standard is being met, you can take your case to the Private Rented Housing Panel (PRHP).

Private Rented Housing Panel

The PRHP is an independent body supported by the Scottish Courts and Tribunals Service which provides a mediation service in repairing obligation cases.

If mediation is not successful, an inspection of your home may take place and a hearing held. It may decide:
- whether your landlord has failed to comply with the Repairing Standard or not;
- to issue an enforcement notice if your landlord has failed to comply, setting out the work to be completed;
- to reduce your rent during some of the enforcement order period.

If an enforcement notice is issued, it sets out the repair work required, and when it must be completed (this will be at least 21 days). If your landlord fails to comply with the notice, the local authority may undertake the work (and charge the landlord).

Repairs and improvements

For England and Wales generally, and in Scotland in connection with the following rights, it is important to distinguish between 'repairs' and 'improvements'. If the works which are needed constitute improvements, rather than just repairs (and the case cannot be brought under the headings of 'negligence' or 'premises prejudicial to health' – see p207 and p208), a landlord has no obligation to improve a home, and you have no rights.[102]

Local authorities have a duty to control premises prejudicial to health under public health legislation.[103] Report your situation to your local authority's environmental health department which can give you advice and take enforcement action if necessary.

You may also take legal action against your landlord for disrepair only if s/he knows about it or should have known about it.[104] It is best to tell your landlord in writing (keeping copies) about the disrepair so that there can be no dispute about whether notice has been given.

Your rights are set out in the Landlord and Tenant Act 1985 or the Housing (Scotland) Act 1987 and the Housing (Scotland) Act 2006. (These provisions do not apply to tenancies for a fixed period of seven years or more.)

Structure and exterior

Your landlord must keep in repair 'the structure and exterior of the dwelling-house (including drains, gutters and external pipes)'.[105] This includes walls, roofs, windows and doors. If these are not kept in good repair, a house can become damp and hard to heat. In Scotland, the property must not be 'below tolerable standard' (see 190).

Installations for heating and for the supply of gas and electricity

Your landlord must keep in repair and proper working order installations for space heating (ie, central heating, gas and electric fires), for heating water and for the supply of gas and electricity.[106] This does not include fittings or appliances making use of the supply – ie, wiring and pipes are included but not cookers or refrigerators. For tenancies which started after 15 January 1989 (2 January 1989 in Scotland), a central heating boiler in the basement of a block of flats would normally come within the repairing obligation.[107]

What you can do

If your landlord does not keep the structure, etc in good repair, you have two options.

- You can bring an action for damages and for a court order requiring your landlord to carry out the repair. Damages are calculated by assessing how much the value of the premises to you has been reduced so as to put you, as far as possible, in the same position as if there had been no breach.[108] This may involve calculating the costs of alternative accommodation, redecoration, eating out, using public baths or launderettes, together with an amount for discomfort and inconvenience arising from the disrepair. Keep a record, as far as possible, of all expenses. Most claims are made in the county court or, in Scotland, the sheriff court. You will need the help of a solicitor. In Scotland, you can take your case to the PRHP (see p205).
- In some cases, rather than taking your landlord to court, it is easier to do the repair work yourself and recover the costs by withholding rent to the same

value. Always write to your landlord to warn her/him of what you are doing. You cannot recover the costs unless the works fall within your landlord's repairing obligations, so you must give her/him an opportunity to object or comment. Send estimates for the cost of the work to your landlord and give her/him time to comment on what is being suggested – eg, 21 days. After the work has been done, write to your landlord to warn that, unless s/he pays the costs, rent to the same value will be withheld. These costs are a 'set-off' against rent due and are not treated by a court as rent arrears, provided the court agrees that the costs were reasonable.[109] The consequences of getting this procedure wrong can be serious, so get legal advice. Note that in Scotland, this way of retaining your rent is not available if you are a statutory tenant.

Negligence

As a tenant, you can hold builders,[110] developers,[111] architects and building engineers[112] liable for their work in building or developing your home if that work was carried out negligently[113] and it causes damage to you or your property or belongings.[114] Local authorities may also be liable for negligence if they fail to inspect properly the plans for, or the site of, your home or to enforce the appropriate building regulations.[115]

If a landlord has repairing obligations (see p203), s/he is also under a duty to make sure that anyone else who could be expected to be in the premises will not suffer harm from any disrepair.[116] Effectively, this extends the repairing obligations to your guests and members of your family, such as your children, even though they are not parties to the tenancy itself. In England and Wales, this duty is specifically extended to situations where the works are carried out before a tenancy is granted.[117] Your landlord is treated as having such repairing obligations if s/he has reserved the right to enter your home to carry out any maintenance or works of repair.[118] Unlike the repairing obligations set out in the previous section, your landlord can be liable under this duty in England and Wales not only if you have given her/him notice of any problem, but also if s/he ought to have known about it.[119]

Electricity and gas can be dangerous. You are protected by safety regulations which prescribe standards and methods of installation of meters and other equipment for the supply of gas or electricity.[120] If landlords carry out work on the premises, they must comply with such standards and are also under a duty to use reasonable care to ensure the safety of those who might be affected by the work.[121]

Failure to meet the appropriate standards may be negligence. The main remedy for negligence is to claim damages in a court action. These are assessed so as to put you, as far as possible, in the position you would have been in had there been no negligence. Legal advice is essential.

Landlords are bound by the Gas Safety (Installation and Use) Regulations 1998 to maintain in a safe condition gas fittings and flues which serve any relevant gas fitting[122] and keep records for two years.[123] A landlord is required to supply a copy of any gas safety record which may be required by the Gas Safety (Installation and Use) Regulations 1998 with the tenancy details and agreement.[124]

Premises prejudicial to health

The Environmental Protection Act 1990 gives a remedy to any person 'aggrieved' by a 'statutory nuisance'. The Act defines 'statutory nuisance' to cover a range of matters. For people who live in defective premises, the most relevant of these is 'any premises in such a state as to be prejudicial to health or a nuisance'.[125]

Severe damp, including condensation, is generally accepted as being prejudicial to health for the purposes of the Act. Loose or exposed wiring and draughty windows and doors are other examples.[126] Health is distinguished from accidental physical injury and would cover, for instance, health problems triggered by gas leakage. There may be grey areas such as cracked electrical fittings which might result in electrical shocks.

A '**nuisance**' is anything coming from neighbouring property which causes substantial interference with your use and enjoyment of your home.[127]

Local authorities have a duty to investigate complaints of statutory nuisance. The local authority may serve a notice requiring any 'nuisance' to be 'abated' – ie, put right. If the notice is not appealed against or complied with, the local authority can prosecute the person who was sent the notice and/or do the works itself. Where the offending landlord is the council itself, it may also be prosecuted under these provisions by you taking a private prosecution.

You can take your landlord to the magistrates' court or, in Scotland, the sheriff court.[128] Legal advice should be obtained. You must give 21 days' written warning to your landlord that you are going to take proceedings. You then 'lay an information' at your local magistrates' court giving details of the defective premises and why they are prejudicial to your health and/or that of any other occupier of the premises. In Scotland, the procedure is by 'summary application' at the local sheriff court.[129] At the subsequent hearing, you must prove the existence of the statutory nuisance and that your landlord is responsible. Environmental health officers can give evidence of the existence of a statutory nuisance. Expert evidence on the state of premises is sufficient to find that premises are prejudicial to health, without having to prove that you are suffering from a condition. A doctor's report explaining the danger to health may be used, and in some cases it may be useful to call the medical practitioner or other expert as a witness.

The proceedings in the magistrates' court follow the rules for criminal procedure, and a finding that a statutory nuisance exists ranks as a criminal conviction[130] – an outcome which most landlords will wish to avoid.

The court can make an order that your landlord must 'abate' the nuisance. The court has wide discretion over what work it may order a landlord to do,[131] although it must be for abating the nuisance. As explained above, repairing obligations can be limited, so this kind of action can be useful if something additional, including improvements, is needed – in some cases courts have ordered the installation of central heating, double glazing and mechanical ventilators. Where a person contravenes any requirement or prohibition imposed by an order, a fine of up to £5,000 may be imposed, together with a fine at a rate of £200 a day for each day on which the offence continues after conviction.[132]

In England and Wales, the magistrates' court can make a compensation order.[133] The order can be for up to £5,000 for things such as damaged belongings and discomfort and inconvenience, although only if the loss was suffered after you sent the 21-day notice.[134] If a court refuses to make a compensation order, it must give reasons. In Scotland, the proceedings are civil, not criminal, and the court has no power to award compensation.

There is a small risk that you might have to pay the defendant's legal costs if you lose,[135] so take legal advice before starting a prosecution. However, so long as the statutory nuisance existed at the time you started the court proceedings, you can ask for your reasonable costs to be paid by your landlord.[136] Also, lawyers can represent you in court on the basis that they will only be paid if the case is successful.[137] Therefore, although financial assistance is not available, if you can find a lawyer who will take the case on such a 'no win, no fee' basis, then it need not cost you anything.

Condensation

Condensation dampness causes severe problems for many people, particularly those living in post-war system-built flats. The dampness and consequent mould growth can be damaging to health and can destroy clothing and furnishings. Attempts to heat damp premises can lead to high fuel bills. The causes of, and remedies for, condensation are complex. Most remedies are beyond the means or control of tenants, involving substantial expenditure on, for example, structure and heating systems.

Legal remedies for condensation

In Scotland, the obligations on a landlord under the Repairing Standard (see p204) are wide enough to cover condensation. This means your landlord has to make sure that there is no condensation problem when your tenancy starts and that, if it arises during the tenancy and you report it, s/he must carry out whatever works are necessary to solve the problem.

In England and Wales, for condensation to come within a landlord's repairing obligations you must show that there has been 'damage to the structure and exterior which requires to be made good'.[138] This has to relate to the physical condition of the structure or exterior. Unless condensation has occurred over a long time and plaster has perished or window frames are rotten as a result, it may be hard to show this.

If the condensation damage is caused by inherent defects in the building (eg, because of defective materials) and if the only way to correct this is to carry out improvements, this can be ordered by the court. A landlord will not, however, be ordered to renew a building completely or to change it substantially – what will be required is a question of degree.[139] It is very unlikely that a court would order installation of a different heating system or the full range of works necessary to remedy condensation.

Therefore, in England and Wales, it is normally more effective to prosecute under the Environmental Protection Act 1990 for a 'statutory nuisance' (see p208). It is not necessary to prove a breach of any contractual or statutory duty to use this remedy.[140] This means that a court can hold a landlord liable even if s/he is not in breach of her/his responsibilities for repairs. A court can also order works of improvement if these are necessary to abate a nuisance.[141]

Landlords sometimes argue that tenants could avoid the nuisance by changing their lifestyle or by heating premises properly. This is rarely correct. If your landlord provides ventilation or a heating system, you are expected to use it,[142] but you are not required to use 'wholly abnormal quantities of fuel'.[143]

In a private civil claim arising from nuisance caused by another tenant rather than the premises, the landlord is only liable to the extent that s/he 'must either participate directly in the commission of the nuisance' or has 'authorised it by letting the property'. Merely being aware of a tenant creating a nuisance is not sufficient to ground a claim.[144]

Other local authority powers

Local authorities have powers to bring unfit properties up to certain minimum standards. Compared with your rights, these powers are more detailed and wide-ranging. However, the disadvantage is that you have to rely on a local authority's willingness to use its powers, which can be limited, mainly due to budget restrictions. Such financial limitations are particularly relevant where mandatory grants are available to bring homes up to the relevant standard (see Chapter 12).

In England and Wales, the relevant standard is prescribed by the Housing Health and Safety Rating System under Part I of the Housing Act 2004. Exactly what satisfies each of these standards is further defined in government guidance (see p189).

In Scotland, the relevant standard is the 'tolerable standard', which is less comprehensive.[145]

Once a property has been identified as falling below the relevant standard, local councils have duties to inspect and make arrangements for dealing with it.

If you feel that your home falls below the relevant standard, contact your local authority to urge it to take action.

Maintenance of gas appliances

Landlords often provide gas fires and other gas-fired appliances. Your landlord must maintain any such appliance or installation pipework owned by her/him in a safe condition so as to prevent risk of injury to any person.[146] Your landlord also has to ensure that each appliance is checked at least every 12 months by a registered gas engineer.[147] Your landlord must give you a copy of the gas safety record within 28 days of it being carried out or before you move in. If you have a social landlord, it is a term of your tenancy that you allow the landlord access to perform necessary safety checks and you risk court action if you prevent or obstruct entry.

Gas Safe Register is the statutory gas registration body in Great Britain. It lists all gas engineers who are qualified to work legally and safely on gas appliances. By law all qualified gas engineers must be on the Gas Safe Register and carry a photo ID card with their licence number and the type of work they are qualified to do. The register is at www.gassaferegister.co.uk or you can call 0800 408 5500.

Using the Ombudsman

Where your landlord is a registered social landlord (ie, a local authority or a housing association) which fails to carry out its duties and obligations, you can make a complaint to the **Housing Ombudsman**. If you are a private tenant, check to see if your landlord is a voluntary member of the Ombudsman scheme.

The Housing Ombudsman provides a free and impartial service investigating complaints brought against social landlords and may recommend that an award of compensation is paid. You must exhaust all stages of the social landlord's complaints procedure before contacting the Ombudsman. The Housing Ombudsman awards compensation, enforceable through the courts.[148] For more information, see www.housing-ombudsman.org.uk and Chapter 14.

In Wales, you can take up social housing complaints with the **Public Services Ombudsman for Wales** (see www.ombudsman-wales.org.uk).

There is also a **Property Ombudsman**. This may be applicable where estate agents are involved with lettings on behalf of landlords, particularly those based overseas. The Property Ombudsman scheme is a free, non-statutory scheme. For more information, see www.tpos.co.uk.

Scottish Housing Regulator

In Scotland, complaints may be investigated by the Scottish Housing Regulator which has jurisdiction to cover complaints made by tenants involving local authorities and registered social landlords. For details, see www.scottishhousingregulator.gov.uk.

5. **Energy efficiency measures**

Private rented tenants in England and Wales

If you are a private renter, you usually require your landlord's consent to carry out energy efficiency improvements to the property. The Energy Efficiency (Private Rented Property) (England and Wales) Regulations 2015 sets out a statutory scheme to make it easier to obtain the required permission from your landlord for specified improvements. Details and guidance can be found in *Private Sector Tenants' Energy Efficiency Improvements* published by the Department of Energy and Climate Change (part of the Department for Business, Energy and Industrial Strategy from July 2016).

You are entitled to apply if you rent your home under:

- an assured tenancy under the Housing Act 1988;
- a regulated tenancy defined under the Rent Act 1977;
- certain agricultural tenancies.[149]

Two or more tenants can make a request together if occupying accommodation in the same building.

Arranging the funding is your responsibility, not your landlord's. The relevant funding sources are:

- the Energy Company Obligation (see p166);
- local authority or central government funding or a third party funding such as a grant (see Chapter 12);
- a Green Deal finance plan (or future equivalent). **Note:** government funding for the Green Deal has ended but you may still apply for grants from other sources. Contact the Energy Saving Advice Service on 0300 123 1234 for details of the remaining Green Deal schemes;
- self-funding. You can pay the full cost or top up funding from the sources listed above.

The energy efficiency improvements you can request are:

- any measures from a prescribed list in the Schedule to the Green Deal (Qualifying Energy Improvements) Order 2012 (see Appendix 5).[150] These include boilers, hot water and heating systems, glazing, draught proofing, loft, rafter and wall insulation, under-floor heating and solar panels;
- the installation of pipes to connect to the gas grid if the property is within 23 metres of a relevant gas main.[151]

Your landlord (and, in a case of sub-letting, a superior landlord whose consent is required) must not unreasonably refuse consent to the requested improvements unless an exemption applies or the landlord proposes alternative energy efficiency measures.

Tenants' energy efficiency improvements request process

Step one: consider the energy efficiency measure you would like installed and how you would fund the cost of the measure and installation. Check if you can get help from the schemes detailed in Chapter 12.

Step two: make a formal written request to your landlord to ask for permission for the measure. You must state:[152]

– the measure you wish to install;

– the works you will carry out (eg, redecoration) to ensure the property is returned to its original condition after the installation;

– details of a Green Deal plan, if relevant.

You must include relevant documents about the measure (eg, an Energy Performance Certificate (EPC) or Green Deal recommendation) and evidence of funding or details of how it may be achieved free of charge. Where the consent of another tenant or third party in the property is needed a copy of that consent should be included.

Step three: your landlord must then consider the request and obtain any further advice, evidence or consents needed. S/he should respond to you within one month to consent, refuse or offer a counter proposal.[153]

Step four: consider your landlord's response, whether you wish to accept it or whether you wish to renegotiate. You can appeal to the First-tier Tribunal if you think the landlord has not complied with the regulations (see below).

Once the measure is installed, it becomes part of the property's fixtures and fittings and so you cannot remove it at the end of your tenancy unless this is agreed with the landlord at the time the consent is granted.

Challenging a decision

If you are not satisfied that your landlord has complied with the Regulations, there is a right to an appeal to the First-tier Tribunal (General Regulatory Chamber).[154] It is recommended that you initially try to resolve the matter direct with your landlord, and seek advice before progressing to the tribunal.

You can apply to the First-tier Tribunal within 28 days from the date of your landlord's response. If your landlord has not responded, this is 28 days from the last date s/he should have replied by. Note that there is a charge of £100 to appeal. Grounds for appeal include the landlord:

- failing to respond to your request;
- refusing consent in breach of the Regulations;
- failing to install measures in accordance with her/his counter proposal;
- serving a non-compliant counter proposal.

Future changes

From October 2017, local authorities (known as 'enforcement authorities') will have powers to enforce provisions on energy efficiency where a property falls below the energy efficiency standard. A privately rented property will be classed as 'sub-standard' where the valid EPC indicates the energy performance of the property is below the minimum level of energy efficiency.[155] From 1 April 2018, the enforcement authority may issue a financial penalty notice against a landlord who has been in breach of regulations and a publication penalty (which consists of publishing the details of the breach in the register maintained by the Secretary of State).[156]

Notes

1. You and your landlord
1 s33 DA 2015
2 s33(4) DA 2015
3 *Leisure Employment Services Ltd v Revenue and Customs Commissioners* [2007] EWCA Civ 92
4 Reg 6(2)(a) EPB(EW) Regs
5 Reg 9 EPB(EW) Regs
6 Reg 13 EPB(EW) Regs
7 Reg 3 ASTNPR(E) Regs

2. Rent increases for fuel or fuel-related services
8 *Montague v Browning* [1954] 2 All ER 601
9 Part 1 and Sch 1 HA 1988; Form 1 Assured Tenancies and Agricultural Occupancies (Forms) (England) Regulations 2015 No.620
10 Part IV HA 1985; s160 LA 2011
11 *Associated Provincial Picture Houses v Wednesbury Corporation* [1948] 1 KB 223
12 s108 HA 1985
13 s211 H(S)A 1987
14 s14(2) Supply of Goods and Services Act 1982
15 s11 LG(MP)A 1976; Sale of Electricity by Local Authorities (England and Wales) Regs 2010 No.1910; Sale of Electricity by Local Authorities (Scotland) Regs 2010 No.1908
16 s12(4) LG(MP)A 1976
17 *Bromley LBC v GLC* [1982] 1 All ER 129

18 Part III London County Council (General Powers) Act 1949
19 s22 London County Council (General Powers) Act 1949
20 s20(3) London County Council (General Powers) Act 1949
21 *R (on the application of Ofogba) v Secretary of State for Energy and Climate Change* [2014] EWHC 2665 (Admin)
22 *Camden London Borough Council v Leaseholders of 46 flats in Harben Road Estate* [2015] LON/00AG/LDC/2014/0123, 27 April 2015
23 *Camden London Borough Council v Leaseholders of 46 flats in Harben Road Estate* [2015] LON/00AG/LDC/2014/0123, 27 April 2015
24 DoE Circular 6/90, Local Government and Housing Act 1989, Area Renewal, Unfitness, Slum Clearance and Enforcement Action, Annex A, Guidance Notes on the Standard of Fitness for Human Habitation
25 ss18–30 LTA 1985
26 ss47–51 HA 1985
27 s18 LTA 1985
28 s26 LTA 1985
29 s27 LTA 1985
30 s19 LTA 1985
31 See *Russell v Laimond Properties Ltd* (1983) 269 EG 947; *Levitt and another v London Borough of Camden* [2011] UKUT 366 (LC)
32 s20B LTA 1985

33 s20B(1) and (2) LTA 1985; *Brent London Borough Council v Shulem B Association Ltd* [2011] EWHC 1663; *Gilje v Charlgrove Securities* [2004] 1 All ER 91
34 [2013] EWCA Civ 479
35 *Westleigh Properties v Grimes* [2014] UKUT 213 (LC)
36 *Waverley Borough Council v Arya* [2013] UKUT 501 (LC)
37 Sch 1 para 29 Mobile Homes Act 1983; *Britanniacrest v Bamborough and Another* [2016] UKUT 144
38 s19(2A) and (2B) LTA 1985
39 Reg 26 The Tribunal Procedure (First-tier Tribunal) (Property Chamber) Rules 2013 No.1169
40 s21 LTA 1985 as amended by Sch 12 Housing and Regeneration Act 2008; Housing and Regeneration Act 2008 (Commencement No 2 and Transitory Provisions) Order 2008 No.3068
41 s21(6) LTA 1985
42 s38 LTA 1985; 'dwelling' is defined as a building or part of a building occupied as a separate dwelling. Provided the occupants of a house in multiple occupation are tenants with exclusive occupation of at least a room, their landlord would have to provide certified accounts.
43 s22 LTA 1985
44 s29 LTA 1985
45 s19(4) LTA 1985
46 ss112 and 176 Commonhold and Leasehold Reform Act 2002
47 Sch 11 para 5A Commonhold and Leasehold Reform Act 2002
48 *Finchbourne Ltd v Rodrigues* [1976] 3 All ER 581
49 See ss18-24 Private Housing (Tenancies) Scotland Act 2016
50 *Otter v Norman* [1989] AC 129
51 s51 RA 1977; s34 R(S)A 1984
52 s47 RA 1977; s31 R(S)A 1984
53 s47(2) RA 1977; s31(2) R(S)A 1984
54 s67 RA 1977; s46 R(S)A 1984
55 s71(1) RA 1977; s49(1) R(S)A 1984
56 s72A RA 1977
57 Sch 1 paras 4 and 5 HB Regs; reg 2 Rent Officers (Housing Benefit and Universal Credit Functions) (Local Housing Allowance Amendments) Order 2013 No.2978 amending the Rent Officers (Housing Benefit Functions) Order 1997 and the Rent Officers (Housing Benefit Functions) (Scotland) Order 1997
58 *Metropolitan Properties Co Ltd v Noble* [1968] 2 All ER 313
59 s71(4) RA 1977; s49(6) R(S)A 1984
60 *Oxford City Council v Basey* [2012] EWCA 115
61 s77(1) RA 1977 ; s72 HA 1980; s65(1) R(S)A 1984 with functions transferred under the Transfer of Tribunal Functions Order 2013 No.1036
62 s78(2) RA 1977; s66(1) R(S)A 1984
63 s80(1) RA 1977; s68 R(S)A 1984

3. **Resale of fuel by a landlord**
64 s44 EA 1989
65 Ofgem, *Maximum Resale Price Provisions*, January 2002, paras 2.8-2.9
66 *The resale of gas and electricity: guidance for resellers*, Ofgem, October 2005
67 *The resale of gas and electricity: guidance for resellers*, Ofgem, October 2005
68 para 3 *Direction made under s37 of the GA 1986 as to maximum prices for reselling gas*, Ofgas, 15 February 1996
69 Ofgem, *Decision on the application of the Maximum Resale Price to the resale of electricity for charging electric vehicles,* 14 March 2014
70 Meters (Approval of Pattern or Construction and Manner of Installation) Regulations 1998 No.1565; Meters (Certification) Regulations 1998 No.1566
71 s17 GA 1986
72 s97 HA 1985; s57 H(S)A 1987; s81 HA 1980; s101 R(S)A 1984
73 Sch 2B para 7(3) GA 1986
74 Sch 6 para 1(6) EA 1989
75 See *Perera v Vandiyar* [1953] 1 All ER 1109; *McCall v Abelesz* [1976] 1 QB 585 Court of Appeal
76 Per Lord Denning in *McCall v Abelesz* [1986] 1 QB 585
77 *Watts v Morrow* [1991] 4 All ER 937; *Halcyon House v Baines and others* [2014] EWHC 2216 July 14
78 *Moorjani v Durban Estates Ltd* [2015] EWCA (Civ) 1252
79 ss233 and 372 Insolvency Act 1986
80 Insolvency Service, *Technical Manual*, Chapter 33, Part 11
81 ss1 and 3 Protection from Eviction Act 1977; R(S)A 1984 as amended by s38 H(S)A 1988; see also *R v Sakaut (Sajeed)* [2014] 0548/B4
82 s27 HA 1988; s36 H(S)A 1988
83 *Rookes v Barnard* [1964] AC 1129 at 1227; *Stratton and Anr v Patel and Another* [2014] EWHC 2677
84 *Kenny v Preen* [1963] 1 QB 499 at 512, CA

85 s19(8) Greater London Council (General Powers) Act 1972; s33(5) LG(MP)A 1976

86 s33 LG(MP)A 1976

87 s19 Greater London Council (General Powers) Act 1972 as amended by s42 London Local Authorities Act 1990

88 s19(3)(b) Greater London Council (General Powers) Act 1972

89 Registering a charge means to attach a charge to the title of the property at the Land Registry so that the owner cannot sell the property without paying off the charge.

90 HA 1985; H(S)A 1987

91 *Gas Competition and your Tenants: a guide for social landlords*, Ofgas Bulletin for Social Landlords, No.1, April 1998

4. Defective housing and heating systems: tenants' rights

92 ss33-34 DA 2015

93 s21 HA 1988

94 ss53 and 54 H(S)A 1987

95 For example, s11 LTA 1985

96 *Joyce v Liverpool City Council; Wynne v Liverpool City Council* [1995] 3 All ER 110

97 www.justice.gov.uk/courts/procedure-rules/civil/protocol/prot_hou#IDA1CKCC

98 Erskine's Institutes II/4/63

99 s27 H(S)A 2001

100 s13 H(S)A 2006

101 s12 H(S)A 2006

102 *Ravenseft Properties Ltd v Davstone Holdings Ltd* [1979] 1 All ER 929

103 See Environmental Protection Act 1990; Public Health Act 1936

104 *O'Brien v Robinson* [1973] AC 912

105 s11(1)(a) LTA 1985; Sch 10 para 3(1)(a) H(S)A 1987

106 s11(1)(b) and (c) LTA 1985; Sch 10 para 3(1)(b) H(S)A 1987

107 s11(1A) LTA 1985; Sch 10 para 3(1A) H(S)A 1987

108 *Calabar Properties v Stitcher* [1984] 1 WLR 287

109 *Lee-Parker v Izzett* [1971] 1 WLR 1688; *British Anzani (Felixstowe) Ltd v International Marine Management (UK) Ltd* [1980] QB 137

110 *Gallagher v McDowell Ltd* [1961] NI 26

111 *Batty v Metropolitan Property Realisations* [1978] QB 554

112 *Cedar Transport Group v First Wyvern Property Trustees Co* [1981] EG 1077

113 s1 DPA 1972

114 *Murphy v Brentwood DC* [1990] 3 WLR 414

115 *Murphy v Brentwood DC* [1990] 3 WLR 414

116 s4 DPA 1972; s3 Occupiers' Liability (Scotland) Act 1960

117 s3 DPA 1972

118 s4(4) DPA 1972

119 s4(2) DPA 1972

120 GS(IU) Regs; Meters (Approval of Pattern or Construction and Manner of Installation) Regulations 1998 No.1565; Meters (Certification) Regulations 1998 No.1566 as amended by 2002 SI 3129

121 *AC Billings & Son v Riden* [1957] 3 All ER 1

122 s30A HA 1988

123 Reg 36(3)(c) Gas Safety (Installation and Use) Regulations 1998 No.2451

124 Reg 36(2) Gas Safety (Installation and Use) Regulations 1998 No.2451

125 s79(1)(a) Environmental Protection Act 1990

126 But see *R v Bristol City Council ex parte Everett* 13 May 1998 – a dangerous staircase is not a statutory nuisance

127 *National Coal Board v Neath BC* [1976] 1 WLR 543

128 s82 Environmental Protection Act 1990

129 Sheriff Court Summary Application Rules 1993 SI 1993/3240 r.4

130 *Herbert v Lambeth LBC* (1991) *The Times*, 21 November 1991

131 *Whittaker v Derby Urban Sanitary Authority* [1885] LJMC 8

132 s82(8) Environmental Protection Act 1990

133 s35 Powers of Criminal Courts Act 1973

134 *R v Liverpool Crown Court ex parte Cooke* [1996] 4 All ER 589

135 s18 Prosecution of Offences Act 1985

136 s82(12) Environmental Protection Act 1990

137 *Thai Trading v Taylor* [1998] *The Times*, 6 March 1998

138 Dillon LJ in *Quick v Taff Ely BC* [1985] 18 HLR 66

139 *Ravenseft Properties Ltd v Davstone Holdings Ltd* [1979] 1 All ER 929

140 *Birmingham DC v Kelly* [1985] 17 HLR 572

141 *Birmingham DC v Kelly* [1985] 17 HLR 572

142 *Dover DC v Farrar* [1980] 2 HLR 32

143 *GLC v LB Tower Hamlets* [1983] 15 HLR 54

144 *Southwark London Borough Council v
Mills* [2001] 1 AC 1; *Lawrence and
another v Fen Tigers Ltd and others* [2014]
2 All ER 622
145 s14 H(S)A 1987; see
www.scotland.gov.uk/Publications/
2003/09/18167/26257 for guidance
146 Reg 36(2) GS(IU) Regs
147 Reg 36(3)(a) GS(IU) Regs
148 ss181 and 182 LA 2011

5. **Energy efficiency measures**
149 Under reg 2 Energy Efficiency (Domestic
Private Rented Property) Order 2015
150 The Green Deal (Qualifying Energy
Improvements) Order 2012 No.2105;
reg 6 (1)(a) EE(PRP) Regs
151 Reg 2(1) EE(PRP) Regs
152 Regs 3 and 8 EE(PRP)(EW) Regs
153 Reg 12 EE(PRP)(EW) Regs
154 Reg 17 EE(PRP)(EW) Regs
155 Reg 22(a) EE(PRP)(EW) Regs
156 Reg 36 EE(PRP)(EW) Regs

Chapter 14

Remedies

This chapter covers:

1. Available remedies

This chapter rounds-up the remedies available if you are in dispute with a supplier of electricity, or a supplier or transporter of gas. The ultimate arbiters of such disputes are Ofgem and the Energy Ombudsman or the civil courts which determine and enforce the standards and protections given by law. The courts may be used by both official bodies and individual consumers.

Under European law the UK is required to ensure the integrity and transparency of the energy market[1] and have national authorities with investigatory and regulatory enforcement powers. The responsible regulatory authority is Ofgem, with other functions exercised by other state bodies or state-supported organisations. Ofgem is an independent regulator, not acting on behalf of you or the supplier, so you cannot instruct it what to do.

European law requires 'that customers benefit through efficient functioning of their national market, promoting effective competition and helping ensure consumer protection' and, acts in 'helping to achieve high standards of universal and public service in electricity supply, contributing to the protection of vulnerable customers'.[2] With the referendum vote in 2016 for the UK to leave the European Union, it remains unclear to what extent these obligations will be recognised within UK law, if at all, once succession occurs. However, an immediate impact was recorded in an increase in energy prices, and issues arise over the long-term maintenance of parts of the energy infrastructure which are neither owned or controlled by UK-based suppliers.

In the UK, the protection of consumer interests was previously the responsibility of Consumer Focus/Futures which closed on 1 April 2014. A number of its services and functions have been transferred to Citizens Advice consumer service.

The statutory remedies described in this chapter have limitations. For all practical purposes, taking your own legal action is sometimes a better way of asserting your rights. However, it costs time and money and, because of restrictions in legal aid, there may be no realistic prospect of receiving help with any civil claim beyond £10,000 except in a small number of cases.

The first step in any dispute is to approach the supplier or transporter and attempt to negotiate with it, being prepared to make a complaint if necessary.

Although not strictly a remedy, the possibility of media attention and publicity in the press, on television or via social media should not be overlooked.

2. **Negotiations**

Negotiating with the supplier or transporter can be the most appropriate way of resolving a problem or dispute, with the back-up of a civil claim through the courts if it is not resolved. To negotiate effectively, you need to rely on a range of documents which, in their different ways, provide 'rules' about how suppliers and transporters should behave. Chapter 1 gives background information on sources of law and it may be useful to read it first, together with any information and guidance originally issued by the National Consumer Council from 2008.

If you contact the supplier, keep a record of the person/section you contacted. Putting your complaint in writing is preferable (keep a copy of all correspondence) and essential if you wish to pursue a complaint. It is also important to keep these records should you decide to pursue a civil claim through the courts.

Using codes of practice and policy statements

Suppliers are subject to regulations which govern how they handle a complaint. The Gas and Electricity (Consumer Complaints Handling Standards) Regulations set down standards for the handling of complaints and the supply of information to consumers. Every supplier has to conform to the regulations; it should have a code of practice based upon the regulations for handling complaints. It must provide a copy of its complaints handling procedure, free of charge, to any person who requests a copy and must have its complaints procedure in a prominent position on its website.[3] Ofgem reviews the complaints system every two years. Suppliers which breach these regulations may be issued with a penalty notice requiring them to pay compensation. For example, in August 2014, EDF Energy was required to pay £3 million in compensation to vulnerable customers for breaching regulations in the way it dealt with customer complaints, arising from long waiting times when customers attempted to contact the company.[4]

Suppliers have also produced other codes of practice, as required by their licences, and staff may be more familiar with these than with the precise provisions of the law. So long as the provisions of a code of practice support your case, it may be easier and more effective to quote these; alternatively, extracts from the Standard Licence Conditions may be quoted where appropriate. Each code of practice has to be approved by Ofgem before being used.[5]

Each supplier should have codes of practice on:
- payment of bills;
- services for elderly or disabled people;
- using fuel efficiently;
- complaints procedures.

Each supplier should also have codes of practice on prepayment meters and site access. Each supplier also produces various documents on its policies and other useful information. It is obliged to publish information regularly about its performance compared with targets set by Ofgem and by itself. Use these if they support your case, but always be cautious – as a summary of the law, they will not always be accurate.

Also, as the supply of fuel is carried out by contract, not by statutory duty, provisions as to unfair contract terms apply. If you believe that a term in your contract is unfair, you can use this to support any negotiations (see p227).

Failure to follow a code of practice or an inadequate code of practice is not automatically negligence or a civil wrong in itself, but it can be evidence in support of such a claim.[6]

Making a complaint

Where you have a problem with the conduct of an energy company, and you cannot get it resolved or correspondence is ignored, make a complaint. This is important as both Ofgem and the Energy Ombudsman expect you to use the complaints service, if you are capable, before contacting them.

If you have asked an adviser or another person to make the complaint for you, give them a signed authority to act for you, allowing information about you to be released.

Regulations lay down minimum standards for dealing with complaints by suppliers.[7] If making a complaint, you should request a copy of the company complaints policy.

Definition of a complaint
A complaint is any expression of dissatisfaction made to an organisation, related to any one of its products, its services or the manner in which it has dealt with any such expression

of dissatisfaction, where a response is either provided by or on behalf of that organisation at the point at which contact is made or a response is explicitly or implicitly required or expected to be provided thereafter.[8]

This definition is wide enough to include an independent subcontractor used by an energy supplier to carry out certain tasks – eg, the enforcement of warrants of entry and the fitting of prepayment meters. Thus, if a subcontractor behaves wrongly, a complaint can be made under the regulations to the supplier who appointed her/him.

A complaint may be made about any of the following:
- billing – including accuracy of bills, frequency of billing, estimated bills, inaccurate bills, sending bills to the wrong address and issuing bills to the wrong person;
- sales – including misleading sales information and behaviour of sales staff;
- transfers – problems that occur when switching suppliers;
- meters – including faulty meters, inaccurate meter readings and problems with fitting and changing meters;
- prices – increases of prices on agreed contracts, misleading price information, problems with direct debits and credits, payment schemes and lack of notification of increases;
- access – problems with access to low-income schemes, special tariffs and government schemes;
- debt – problems with debt, disconnection and payment of arrears and failure to apply for Fuel Direct where available;
- customer service – inconsistent or inaccurate information, failures by staff, delay in responding to enquiries and problems with prepayment cards.

A supplier is required to have a complaints procedure in place and must comply with it in relation to each complaint it receives.[9] The procedure must:[10]
- be in plain and intelligible language;
- allow for complaints to be made and progressed orally (by phone or in person) or in writing (including email);
- describe the steps it will take to investigate and resolve your complaint and the likely time this will take;
- provide for an internal review of your complaint if you are dissatisfied by the response.

The supplier must also give the names and contact details of the main sources of independent help, advice and information. To be independent, the advisers must not be connected with the energy company.

Your right to refer your complaint to a qualifying redress scheme from the point at which the supplier notifies you in writing that it is unable to resolve your complaint to your satisfaction should be explained.

It is possible for many complaints to be resolved at the initial contact or within a couple of days. The complaints that suppliers are unable to resolve so quickly are more likely to be recorded by the supplier as a complaint. Ofgem focuses on those complaints that remain unresolved by the end of the working day after the complaint has been recorded. Complaint data on companies is available from the 'big six' suppliers since October 2012 and on most of the medium and smaller energy companies since 1 April 2013.[11] This information provides data for potential enforcement action and the imposition of penalties. Ofgem has already taken action over suppliers' complaint handling. In December 2015, Npower had a £26 million penalty imposed as a result of billing and complaint handling failings and in April 2016 Scottish Power had to pay out £18 million for similar failures.

Recording a complaint

On receiving your complaint, a supplier must electronically record that the complaint has been made. It must record the date, whether the complaint was made orally or in writing and your name and contact details or those of the person making the complaint for you.[12]

Where the supplier is licensed by Ofgem, details of the complaint must be recorded along with details of you and your account, together with a summary of any advice given and any agreement on future communication.[13]

Where you have made a complaint but the supplier cannot find your complaint, the supplier must record the fact that it is unable to trace your complaint.[14]

Where a supplier has recorded that your complaint is resolved but subsequent contact from you contradicts this, the supplier must not treat your complaint as a resolved complaint until it is demonstrably a resolved complaint.[15]

If you reach the position where the complaint is not resolved, the supplier must issue a letter saying this. This is known as a 'deadlock letter'. However, in reaching this point, the supplier must also set out the different remedies available to you under the complaints handling procedure, which must include:
- an apology;
- an explanation;
- the taking of appropriate remedial action by the regulated provider; *and*
- the award of compensation in appropriate circumstances.

When a complaint is treated as received

Your complaint and any subsequent communication must be treated as having been received:[16]
- where contact is made orally (by phone or in person), at the time at which it is received by that regulated provider;
- where made in writing (including by email) and it is received:
  - before 5pm on a working day, on that day;

– after 5pm on a working day or at any time on a day that is not a working day, on the first working day immediately following the day upon which it is received.

Where a supplier fails to handle complaints properly, Ofgem may investigate and impose penalties.

Complaints about National Grid

If your complaint is about any aspect of the operation of National Grid or its employees and agents, you can use its free complaints service. National Grid will investigate your complaint and respond to you within 10 working days. If it is not possible to investigate the complaint within the 10 days, it will inform you when a response can be expected. If no response is forthcoming, there may be grounds for compensation under National Grid's standards of service provisions.

3. Action by Ofgem

Enforcement matters

Ofgem has powers to order energy suppliers and gas transporters to do anything it considers necessary to ensure they comply with certain provisions of the Acts or any conditions in their licences. The matters covered by these powers are called **'enforcement matters'** and include:[17]

- giving and continuing to supply electricity;
- connecting premises to a supply of gas;
- paying interest on security deposits;
- keeping meters in proper working order;
- producing codes of practice or other arrangements to deal with customers in default.

The list of enforcement matters seems more limited than it really is. For instance, disputes about responsibility for bills are not specifically mentioned, but may be covered indirectly because one remedy for a supplier in a dispute is to disconnect you and disconnection may be an enforcement matter. Ofgem can intervene in a dispute if it is likely to end up being an enforcement matter, but it cannot intervene in individual billing disputes.

The conditions in suppliers' and transporters' licences are not enforceable by individual consumers because they are obligations arising between the respective supplier and Ofgem. Many disputes arise directly under the relevant Acts, but those that only involve breaches of licence conditions have to be referred to Ofgem. If necessary, Ofgem's exercise of its powers can be judicially reviewed. In certain circumstances, the suppliers themselves may also be judicially reviewed (see p238).

Standards of performance

Failure by suppliers to comply with overall standards of performance can result in regulatory action by Ofgem. Suppliers are regulated through the Standard Licence Conditions (SLCs). If there is widespread evidence of suppliers flouting or avoiding their obligations under the SLCs, Ofgem is expected to act.

There are two kinds of standards of performance – 'overall' and 'individual'. **'Overall standards'** are targets laid down by Ofgem to measure the supplier's general performance. Failure to comply with overall standards is a matter between the supplier and Ofgem and is unlikely to affect you directly. **'Individual standards'** are rules of performance (eg, Electricity (Standards of Performance) Regulations 2015) which, if they are breached, normally entitle you to a small compensation payment.

If you encounter a breach of licence conditions, report the matter to Ofgem for investigation.

Electricity

Standards apply to all electricity suppliers. Suppliers provide online information and leaflets describing them – some may also include additional standards which the supplier has set for itself. The standards set down by the law cover:

- failure of the distributor's fuse;
- restoring supply where disconnection was the supplier's fault;
- providing a supply;
- providing an estimate of charges for connection of a supply or moving a meter;
- giving notice when the supplier has to interrupt a supply;
- dealing with voltage complaints;
- dealing with meter disputes;
- responding to complaints about prepayment meters not working;
- responding to requests or queries about charges or payments;
- making and keeping appointments;
- giving notice to consumers of their rights under this scheme.

The standards require the functions under each of these headings to be carried out within a certain number of working days. Failure to do so entitles you to a fixed sum, from £30 (for most matters) up to £75. Standard payments are normally maximum payments. If failure to meet the standards causes you to lose more than £30 or £75, you can still claim the larger amount. If necessary, you can go to court. If there is any dispute between you and the supplier over these standards or the payments, contact Citizens Advice consumer service for advice (see Appendix 1). Your ultimate remedy would be a claim in the civil courts.

Where a supplier fails to pay you for failure to meet a standard within 10 working days, an additional payment of £30 (known as an 'additional standard payment') must be made.[18]

Gas

Minimum standards are laid down in regulations for gas. As with electricity, there are some automatic levels of compensation for which standard payments and additional standard payments may be made. Gas companies may also have their own schemes of settlement. The customer relations manager in each British Gas region has the authority to settle claims for breach of these standards, up to £5,000.

Obtaining an order from Ofgem

If you think a supplier or transporter may be in breach of an enforcement matter, contact Citizens Advice consumer service in the first instance. If satisfied that there has been a breach, the matter is referred to Ofgem, which makes either a 'provisional order' or 'final order'. Making a provisional order is quicker than a final order, so you should press for the former. Ofgem has powers to obtain compensation on your behalf.

Ofgem cannot make an order if:

- it thinks that its general duties laid down by the Acts do not allow it; *or*
- the breaches in question are trivial; *or*
- it is satisfied that the supplier or transporter has agreed to, and is taking all steps necessary to, comply with its obligations.

Ofgem possesses powers to extend its range of energy industry functions and activities that it may regulate and licence, including the activities of third parties (eg, consumer switching sites) subject to approval by the Secretary of State and a resolution of the House of Commons.[19]

Ofgem must tell you if it decides not to make an order. In deciding whether to make an order, Ofgem must take into account, in particular, your lack of other remedies and the loss or damage which you might suffer during the consultation period, which has to take place before a final order is made. If you would otherwise be without a supply, a provisional order will normally be appropriate.

If a provisional order is made and complied with, Ofgem only confirms it as a final order if further breaches are likely to occur.

Before making a final order or confirming a provisional order, Ofgem must serve you and the supplier/transporter with a copy of the proposed order and allow 28 days for representations. If it wants to modify the original proposal, Ofgem must either get the consent of the supplier/transporter or serve copies and allow a further 28 days. A similar procedure must be gone through to revoke a final order.

Ofgem must also comply with the ordinary legal rules about natural justice which govern public or government organisations (see p238). In one case, a decision by the previous gas regulator Ofgas was quashed by the High Court because it did not tell a consumer that it had interviewed an important witness, nor did it give the consumer a chance to reply to what the witness had said.[20]

You are entitled to a copy of any order when it is made.

A supplier/transporter can appeal to the High Court (Court of Session in Scotland) against the making of an order. Although you would not necessarily be directly involved, you can be added as a third party – this is a technical procedure for which you need to get legal advice.

The supplier has a duty to obey any order. This means you can sue for a breach of statutory duty if Ofgem does not enforce its own order. Ofgem can also enforce its orders by ordinary civil action against the supplier/transporter, which would be easier and cheaper for you, if you cannot obtain legal aid funding to do this yourself (see p230).

Each case is decided on its merits. As mentioned above, cases rarely, if ever, reach this stage because suppliers want to avoid formal (and public) action.

Determinations by Ofgem are final, and once a determination has been made you cannot sue a supplier or transporter over the same matters.

Breach of conditions relating to payment difficulties

All suppliers' licences require them to compile methods to deal with customers in arrears. These matters are dealt with elsewhere in this book, but it is worth pointing out that a supplier's failure to comply with these 'methods' is an enforcement matter because it is a breach of the relevant condition.

Consumer redress orders

Ofgem is empowered to make consumer redress orders which may be used to provide an alternative to lengthy and expensive litigation and benefit all consumers, whether they are aware they have suffered loss or not.[21] An order may require a supplier to provide redress to you directly, compensating you directly where possible or putting you back into the position you were in before the breach.

Ofgem may make a consumer redress order where it is satisfied that a supplier is contravening any condition or requirement and one or more consumers have suffered loss, damage or been caused inconvenience.[22] A loss need not be financial to qualify, so inconvenience or nuisance may be a possible ground.

Under a consumer redress order, a supplier may be ordered to:[23]

- pay compensation to each affected consumer for the loss, damage or contravention;
- issue a written statement setting out the contravention and its consequences;

- terminate or vary any contracts entered into with affected consumers. This can only happen with your consent.

If compensation is part of the order, the amount of compensation must be specified, along with details of the consumer or consumers concerned who will receive it.

Ofgem does not have a prescribed set of penalties; it approaches each complaint on a case-by-case basis to retain the widest discretion.[24] The amount of compensation must be reasonable. The maximum amount of penalty that may be imposed on a regulated supplier may not exceed 10 per cent of its annual turnover. The powers are designed to be proportionate and build upon the power to impose penalties and the role of the Ombudsman. Consumer redress orders may be used together with, or separately from, penalties.[25]

4. Unfair terms and consumer protection

Unfair terms

Consumer Rights Act 2015

As your fuel supply is provided under a contract with the supplier, the law relating to contracts, including regulations covering unfair terms, applies.

The relevant law is found in the Consumer Rights Act 2015 and in decisions by the higher courts. The Act supersedes much of the earlier law[26] but decisions under previous regulations and caselaw continue to be relevant.

An unfair term of a consumer contract is not binding on you.[27] Under the Act, a term qualifies as 'unfair if, contrary to the requirement of good faith, it causes a significant imbalance in the parties' rights and obligations which operate to the detriment of the consumer'[28]. Whether or not a term is unfair is determined by looking at the nature of the contract and all the circumstances existing when the term was agreed, together with all the other terms of the contract.[29]

Rights and remedies under the Consumer Rights Act 2015 are available to individuals (but not small business or companies). Gas and electricity are treated as 'goods' under the Act if supplied and sold in limited volumes or set quantities.

Transparency – that contract terms can be readily understood – is fundamental to fairness. Written terms of a consumer contract, and any written notices to consumers, must be transparent.[30] They must be expressed in plain and intelligible language and be legible. This specific requirement operates alongside the requirement of good faith and open and fair dealing (see p19).

All obligations and rights should be set out fully, and in a way that can be understood by the average consumer so s/he may understand their practical significance.

If a term could have different meanings, the term that is most favourable to you is taken.

Enforcement action for unfair terms may be taken by one of a number of bodies in the UK including the Competition and Markets Authority, the Secretary of State, a district council in England and Ofgem.[31] A number of these bodies may also enforce the provisions of EU law while the UK remains within the EU.[32]

Examples of terms which may be considered unfair[33]

Terms limiting or excluding legal liability for death or personal injury as a result of an act or omission.

Terms which exclude, or limit, your legal rights in relation to the supplier or another party in the event of total or partial non-performance or inadequate performance by the supplier of any of the contractual obligations.

Any terms limiting the option of offsetting a debt owed to a supplier against any claim which you may have against the supplier.

Terms which allow the supplier to retain money you have paid where you decide not to finish the contract, but do not allow you to receive compensation or a refund where the supplier is the party cancelling the contract.

A term that requires you to pay a disproportionately high sum or penalty clause if you decide to opt out of an agreement.

Clauses which seek to limit when you can begin legal proceedings.

Disproportionate penalties and compensation which may be payable if you end the contract.

Provisions that may allow a supplier to unilaterally dissolve or alter a contract without notice or where no such right is given to you.

Clauses which you do not know about when the contract is formed.

Causes allowing the supplier to increase the prices without giving you the right to cancel the contract if the final price is too high in relation to the price agreed.

Clauses which all the supplier to transfer obligations where this may reduce the guarantees for the consumer, without the consumer's agreement.

Consumer Protection from Unfair Trading Regulations 2008

Action may also be taken against a supplier which engages in an unfair trading practice as defined under the Consumer Protection from Unfair Trading Regulations 2008. These cover unfair commercial practices which affect the operation of consumer choice, referred to as a 'transactional decision' in the regulations. A transactional decision has a broad meaning covering any decision taken by a consumer, whether it is to act or to refrain from acting, concerning whether, how and on what terms:

* to purchase, make payment in whole or in part for, retain or dispose of a product; *or*
* to exercise a contractual right in relation to a product.

'Commercial practice' is also given wide meaning and includes a trader's act, omission, course of conduct, representation or commercial communication (eg, advertising and marketing) which is directly connected with the promotion, sale or supply of a product to or from you.[34] The unfair practice can occur before, during or after the transaction, whether or not the transaction ultimately takes place. Thus, unfair attempts to influence you through marketing and cold-calling and steps that might be taken to stop you exercising your rights can be caught by the regulations.

The test of whether a commercial practice is misleading includes whether it contains false information and whether it deceives you into a transaction you would not have otherwise taken.[35] This includes the marketing of a product (including comparative advertising) which creates confusion about any products, trademarks, trade names or other distinguishing marks of a competitor. Importantly, it may also cover the failure by a fuel supplier to comply with a code of conduct if it has indicated that it is bound by the code of conduct, and the breach causes you to enter into a transaction that you otherwise would not have done.

If a trader engages in a commercial practice that is misleading, it is guilty of an offence and may be prosecuted by a local authority's trading standards department.

In 2012, the Court of Appeal upheld the conviction for an offence under these regulations by SSE.[36] The company was held liable for misleading statements made by a trainee salesperson working for a linked company, Southern Electric Gas Ltd, operating in an area with a considerable population of elderly consumers and which had been designated by the local council as a 'no cold calling zone'. The Court of Appeal ruled that both companies could potentially have been prosecuted and that SSE fell within the definition of a 'trader' under the regulations as Southern Electric Gas Ltd was held by it as a subsidiary company.

Consumer protection and advocacy

The 1999 Regulations originally provided for the Office of Fair Trading to pursue cases on behalf of consumers and extended the power to investigate and challenge unfair contract terms in order to allow other bodies to share this enforcement task, including local authority trading standards departments and Ofgem. The power is to challenge for the general protection of consumers, not individuals. The Office of Fair Trading was abolished on 31 March 2014 and a number of its functions relating to consumer advice were transferred to Citizens Advice and Citizens Advice Scotland from 1 April 2014,[37] with the remainder of its functions split between the Competition and Markets Authority and the Financial Conduct Authority (FCA). Complaints about unfair contract terms are now the responsibility of the FCA. The FCA operates a consumer telephone helpline (0800 111 6768).

5. **Using the civil courts**

There are two types of civil court action in the field of fuel rights – ordinary court action and judicial review through the High Court. In England and Wales, the relationship between the consumer and the supplier is based on contract law or the law of tort, with remedies available through the civil courts depending on the amount of harm or damage involved. Judicial review is used for actions against state bodies and regulatory authorities.

In Scotland, ordinary court action may only be available against a supplier which supplies under a contract. It is likely that a Scottish sheriff court would not allow an action for breach of statutory duty, as that is a type of action which has not previously existed in Scottish law.

In Scotland, a consumer's relationship with a public electricity supplier/ regional electricity company or transporter is statutory, and so any claim is based on a breach of a statutory duty. This means that you have to pursue your remedies through Ofgem. On the other hand, Ofgem is subject to judicial review. The procedure for judicial review is more flexible and easier to use in Scotland than in England and Wales, so you should still have an effective remedy.

Legal funding

Before April 2013, Community Legal Service funding was available for taking court action if you had a strong enough case and qualified on financial grounds. However, there are now major restrictions on these cases and realistically legal aid is only likely if a consumer matter also involves a matter of serious personal injury or a threat to loss of home. Check www.gov.uk/legal-aid/eligibility to see if you are likely to qualify.

In Scotland, the Scottish Legal Aid Board provides civil assistance. Eligibility is based upon disposable income after essential expenses have been paid. For more details, see www.slab.org.uk.

In practice, it may be very hard to find a solicitor or advice and assistance funded by legal aid in many areas. It may also take time for an application for legal aid assistance to be processed.

Emergency legal representation

Assuming a legal aid provider can be found, it may be necessary to apply for emergency legal representation in many cases involving energy problems. An emergency certificate will only be granted if:

- there is a risk to your physical safety or that of any family member or your home;
- there is a significant risk of a miscarriage of justice, unreasonable hardship or irretrievable problems in handling the case;
- there are no other appropriate options available to deal with the risk.

An emergency legal aid certificate lasts for four weeks.

As an alternative, it may be simpler and more appropriate to begin a claim in the county court using the arbitration or small claims process. Most civil claims are unlikely to qualify for legal assistance if the amount claimed falls below £10,000. This means that you may have to act for yourself.

Using the small claims procedure in the county court

England and Wales

Every year thousands of people represent themselves in small claims hearings, though until recently they have not often been used by energy consumers. In England and Wales, small claims are heard in the county court which deals with civil cases where up to £100,000 is involved (claims of £100,000 or more are heard in the High Court).

Where a dispute involves less than £10,000 – as with many consumer matters – it is dealt with under a simplified procedure known as 'arbitration' or 'a small claims hearing', normally taking the form of a hearing in private in chambers – ie, the judge's private room.

All designated money claims in civil cases in England and Wales are issued via Northampton County Court and the administration dealt with through one business centre in Salford. The centre is supported by a dedicated contact team that deals with all telephone queries relating to claims. You are encouraged to begin your claim online, although hard-copy forms may be downloaded or available from local county courts which are designated as hearing centres.

Proceedings may be brought against a supplier which is in breach of contract (eg, overcharging or wrongly withholding a refund) and for any harm or damage it may cause, either by itself or its employees or sub-contractors.

The small claims court is a relatively informal procedure, suitable for people who are not represented by a solicitor. This might be appropriate if an unlawful disconnection has caused you a relatively small loss or your landlord has been charging more than the maximum resale price for gas or electricity.

Other claims might include where you have a dispute with an energy company about the amount you have paid or where there has been a failure of supply which has resulted in damage such as loss of frozen food. Or you may have a dispute about the amount of fuel consumed at your home which you cannot resolve with the supplier.

Neither side can claim legal costs beyond the court fees involved and, as a result, suppliers tend to settle these cases rather than spend money on contesting them which will not be recoverable.

When the claim form is issued, a court fee is normally payable. However, if you are on income support (IS), income-based jobseeker's allowance (JSA), pension credit (PC), income-related employment and support allowance (ESA) or universal credit (UC), you may be exempt from paying any fee on application to the court (see p240). If you are not receiving any of these benefits, you can apply for remission or reduction of a fee if you would otherwise suffer undue financial hardship because of the exceptional circumstances in your case.

If you succeed in your claim, the court fee is added to the amount which the other side has to pay you. Unlike other court proceedings, only limited costs can be reclaimed. This means that, even if you lose, you will not have to pay the other side's own legal representation costs – ie, each side is responsible for its own costs. If the case is settled or discontinued, there may be a full refund of the hearing fee if you notify the court in writing, at least seven days (excluding the date of receipt and date of hearing) before the trial date or start of the trial week.

The claim form requires you to set down the details of your legal claim. Copies of the form are then lodged in court on payment of the fee (unless this is waived) and a copy is sent to the other side (known as the 'defendant'). The issue of the form requires the defendant to either admit the claim or to defend it. In either case, the defendant must reply to the issue of the proceedings. If the defendant does nothing, after 21 days you may be entitled to claim judgment in default. This means you can obtain your judgment without having to argue the case in court, simply because the defendant has failed to reply.

Experience suggests that fuel suppliers rarely contest proceedings in the small claims court, as the cost of sending someone to attend the hearing often exceeds the amount of the money concerned. In some cases involving relatively small sums (eg, less than £500), the supplier may not even contest proceedings. This factor encourages the settlement of a dispute. Small claims may be particularly suited to the recovery of deposits.

Under the rules of civil procedure, each side is entitled to see the written evidence and documents used in a claim before the hearing. Each side is expected to list its documents and to make copies available. Neither side should be taken by surprise by written evidence at the hearing.

In bringing a claim relating to overcharging or a failure to supply for which you have been charged, you should gather together all your energy bills. If you no longer have them, request them in writing from your supplier. Also bring a copy of all correspondence and the contract with the energy company. These documents should be organised in date order.

Look at the agreed price for supply and work out whether the company has charged the correct rate for units over 12 months. Check whether it has charged more. Use April 1 to March 31 as the starting and finishing dates. Look especially at any periods where there may have been overcharging. You will need to produce these documents if a case goes as far as court; prior to any hearing you should also send the other side copies of all documentary evidence on which you intend to rely.

For each year, work out how many units over the limit you have been charged and multiply them first by the higher rate and then by the lower rate. The difference is the amount that you should claim for overcharging.

Check whether the supplier has ever given you an explanation of its charging methods. If not, state that you believe that you have been wrongly charged from when you became a customer, to the present day. Calculate your entire usage over the period and work out how much it would have cost when your supply started. Then work out how much you have actually paid. The difference between the two figures is what you should claim. Such cases may also arise from under-charging leading to the supplier suddenly trying to recover money with a demand for a lump sum.

The rules of court encourage parties to try to settle their cases without recourse to court proceedings – at any stage parties can negotiate and make settlement proposals to each other to avoid litigation.

If the matter goes as far as a small claims court hearing, each side has an opportunity to present her/his case. Any written evidence which is presented should normally be served on the other side well before this final hearing takes place.

Transfer of proceedings

If you become involved with litigation with an energy supplier, it may be important to ensure a transfer of proceedings to your nearest county court (or the High Court if the claim exceeds £100,000). An application may be necessary to the court.

When considering whether to grant a transfer, the court must consider:
- the financial value of the claim and the amount in dispute (if different);
- the convenience of moving to another court;
- the availability of a judge specialising in the type of claim in question;
- whether the facts, legal issues, remedies or procedures involved are simple or complex;
- the importance of the outcome of the claim to the public in general.[38]

Methods of service

The Civil Procedure Rules set out the various methods of service that can be used. The usual method of service is by first class post, with the documents deemed served two days after posting. Service may also be by email, fax or through a document exchange.

Witness statements and all documentary evidence should also be sent to the other side in advance of the hearing. For example, you might wish to call an electrician or meter reader as a witness in a case, in which case it will be necessary to submit a written witness statement of what s/he will say first. Documentary evidence should normally be exchanged between both sides in a case before the hearing, subject to directions by the court. Each side gives its evidence to the court and has an opportunity to question the other (a process known as cross examination). The judge may also ask questions of the parties at the hearing.

Representation

A party to a hearing is also entitled to 'quiet assistance' from a friend to help present her/his case. The friend is entitled to take notes, suggest questions and give quiet advice on the conduct of the case. The friend may be legally qualified but this is not essential. Such assistance is known as having a 'McKenzie friend', derived from the case of *McKenzie v McKenzie*,[39] and courts are generally familiar with the concept. McKenzie friends often assist debtors in debt recovery proceedings in certain courts. The McKenzie friend has no right to address the court but, in practice, the courts may allow a McKenzie friend to address the court if a litigant has difficulties. The right to speak is a discretionary one, and anyone granted the right to speak must not abuse the privilege. In particular, it is crucial that any statements made to the court relate to the facts and points of law in the case and are not directed as a general attack on the energy company and its policies. If the right to a McKenzie friend is abused (eg, by making political or personalised attacks), it may be withdrawn. It is important to be polite at all times.

Advice should be taken, since a favourable verdict in court cannot be guaranteed or a lesser sum might be awarded if the case goes as far as a hearing. The details of any settlement should be in writing and marked as 'full and final settlement'. Before the case begins, a payment into court may be made by a party which, if accepted, will settle the case.

The judgment

A small claims court judgment takes effect like any other judgment of the county court and enforcement action can be taken if the party does not follow the terms of the judgment – eg, in a claim for compensation by paying the money owed or awarded.

Scotland

Scotland also has a small claims procedure for sums up to £3,000 using the sheriff court. Procedure is governed by the small claims rules with standard forms known as summonses to be completed. The forms can be obtained from the sheriff court clerk or downloaded from the Scottish Courts Service website (www.scotcourts.gov.uk). The person bringing the action is known as the 'pursuer' and the person who is being taken to court is known as the 'defender'. The details can be amended before a hearing takes place. It is also possible to apply for time in order to try to reach a settlement to the case. Bringing the case to a temporary halt in this way is known as 'sisting' the case.

If the defender does not respond to proceedings, Form 11 should be completed setting out the order you wish to obtain. If a party fails to appear at a small claims hearing, the sheriff can grant an order, known as a decree. In Scotland, no costs are payable if the claim is under £200, but costs can be awarded up to £150 if the claim is over £200.

Disputes over £10,000

In county court proceedings involving sums of more than £10,000, each party has to pay its own costs, and the unsuccessful party also has to pay the other party's costs. These may be substantial and while individuals can represent themselves, it is advisable to instruct a solicitor. Note that even if you are successful, not every cost incurred is recoverable.

As with small claims, the rules of court encourage parties to settle their cases out of court.

Parties to litigation involving larger sums must ensure that they comply with the Civil Procedure Rules and directions and orders from the court in terms of supplying and filing documents to the other party and the court. Both claims and defences may be struck out by the court or parties subject to sanctions from the court where procedural rules are not followed.[40] Decisions to strike out cases or apply sanctions may be appealed where there is a reason for failure to comply. The Court of Appeal gives an example: 'if the reason why a document was not filed with the court was that the party or his solicitor suffered from a debilitating illness or was involved in an accident, then, depending on the circumstances, that may constitute a good reason.'[41]

Claims for harassment and damages

An important case which indicates that the courts will not tolerate heavy-handed and intimidating actions by energy suppliers was the judgment in *Ferguson v British Gas*[42] where the Court of Appeal ruled that legal threats issued by British Gas could constitute harassment and could be subject to both civil proceedings and a crime under the Prevention of Harassment Act 1997.

In *Ferguson*, over a period of months British Gas sent bills and threatening letters to the claimant who was a former British Gas customer who had switched to npower. The letters demanded money she did not owe. The threats included to disconnect her gas supply, to start legal proceedings and to report her to credit reference agencies. Despite repeatedly contacting British Gas the threats continued, including after she complained to Energywatch and twice to the chairman of British Gas. As a result she wasted many hours, and, more importantly, was brought to a state of considerable anxiety. By January 2007 she had instructed a solicitor but still no response was received. As a consequence she began legal proceedings claiming £5,000 for distress and anxiety and £5,000 for financial loss due to time lost and expenses in dealing with British Gas and that the course of conduct amounted to unlawful harassment contrary to the Protection from Harassment Act 1997.

Excuses raised by British Gas that it could not be blamed for letters issued by a computer or that it was a company, and should be treated as different to an individual who issued threatening letters, were rejected by the court.

The court ruled that a company such as British Gas could be held responsible for mistakes made by its computerised debt recovery system and the personnel responsible for programming and operating it. The company could be held liable in the same way that a human being could be.

The court also indicated that harassment could be a crime as well as a tort or civil wrong and that in 'any well-documented case, what is sufficient for the one purpose is likely to be sufficient for the other'. This ruling opens the way in future for energy companies that allow harassment of debtors to take place to be prosecuted under the Protection from Harassment Act 1997. It was further observed that, the 'primary responsibility should rest upon local public authorities which possess the means and the statutory powers to bring alleged harassers, however impersonal and powerful, before the local justices.' This means that trading standards departments could prosecute where wrongful debt collection turns into harassment.

A wrongful attempt at debt enforcement by an energy supplier can also result in damages for slander or libel if statements are made by a supplier that are untrue and may damage your reputation.[43]

Injunctions/interdicts

An '**injunction**' ('**interdict**' in Scotland) is an order made by a court which either prohibits someone from doing something (a 'prohibitory' injunction) or instructs someone to do something (a 'mandatory' injunction). For example, if a supplier or landlord illegally cuts off your supply, you could ask for an injunction to get it reconnected.[44] Failure to obey an injunction is contempt of court, punishable by fines, or even imprisonment in extreme cases.

In urgent cases you can get an injunction in the absence of the other side (what is known as an *ex parte* injunction) – eg, where locks have been changed without permission or a meter has been unlawfully removed. The injunction is obtained by going to court and making an appointment for an urgent hearing. A standard form is provided, and you are asked to provide evidence in the form of a statement of truth. Normally, the claim for an injunction will be part of an action for damages. The application is made before a judge who makes a decision as to whether the injunction should be granted and gives directions about how the other side (known as the defendant) is to be notified with her/his decision. **Note:** applying for an injunction can be costly. You will need to pay a court fee of at least £155[45] and should also seek advice from a solicitor before using this remedy. The legal help scheme is no longer available for advice about this type of debt, so you would need to meet the cost of instructing a solicitor yourself.

Although there are some situations in which an injunction is granted almost as a matter of course (eg, illegal eviction), you have no 'right' to an injunction. Injunctions are within a court's discretion and whether or not they are granted depends on the overall circumstances of the case.

The most common situation in a dispute with a supplier is where you ask for an 'interim' injunction (or 'interim' interdict in Scotland). This is when you need a temporary court order quickly, usually valid until the whole case can be put properly before the court.

An example where it might be necessary to threaten or seek an injunction is where a supplier starts action to disconnect a supply by mistake – eg, where action is taken against the wrong address. Normally, you will have had a warning but the situation can arise where a supplier or its agent has obtained a warrant against the wrong address and begins steps to disconnect. Or you come home from a holiday to find you have been disconnected in error by warrant concerning another customer.

During normal court opening hours a hearing can usually be obtained very quickly. You should try to give as much notice as possible to the energy company. Email or fax a letter to the supplier where circumstances allow, addressing it to the legal department. The threat of applying for an injunction may be sufficient to obtain a suitable response from the supplier and make applying for the injunction unnecessary.

In this situation, neither you nor the supplier/transporter has time to present your case fully, and the court has to decide without hearing the evidence in full. The court will consider the 'balance of convenience' – ie, whether you or the supplier/transporter has more to lose or gain from the refusal or granting of an interim injunction, including whether a later award of damages will make up for any such loss.[46] For example, where a supplier threatens disconnection, the court will balance the inconvenience to you of being disconnected against the inconvenience to the supplier of having to continue to supply someone regarded as a bad customer. A court almost always considers the balance to be in your favour if you are prepared to agree to a prepayment meter at least until your dispute is resolved.[47] If an immediate injunction is granted you will be asked to serve the supplier with notice of the order immediately by phoning them or faxing a letter; a copy of the order will also be drawn up by the court. Failure to obey an injunction puts the supplier at risk of contempt proceedings, for which it may be fined or individuals may be jailed as a punishment. In emergency situations, injunctions can be obtained outside normal court hours. The court will have a phone number to contact in such cases and injunction can be granted by a judge over the phone.

Damages

As well as, or instead of, an injunction, you can claim damages (ie, monetary compensation) for a supplier or transporter's abuse of its powers or failure to comply with its statutory or contractual duties.[48] If your supply is accidentally cut off, you may be able to claim damages for negligence and damage which has come directly from the interruption of electricity or gas supply (but not for the pure supply interruptions themselves). A claim for damages for breach of contract must be based on the amount necessary to put you in the position you would have enjoyed had the contract been performed. Losses that are reasonably foreseeable at the time the contract was made as a being a likely outcome of the breach of contract may also be recoverable. In a case of negligence and a claim in tort arising from a harmful act, the measure of damages is the sum necessary.

In a case of failure to supply, damages would cover compensation, not only for the distress and discomfort of being without a supply but also for additional expenses (eg, take away meals) and the loss of specific items (eg, fridge/freezer contents). Interest can also be claimed on any sum awarded in damages. In other cases, the failure of a supplier to disconnect a supply may amount to breach of a duty of care where harm results – eg, a fire which arises in circumstances where electricity should have been cut off.[49] All such losses arising from the breach of contract or duty of care by a supplier need to be specifically set out and pleaded in any claim.

A claim for negligence may also arise where a supplier has caused damage or breached safety rules and damage to a person or property has resulted. In *Smith and others v South Eastern Power Networks plc and other cases,* what is safe is judged as an objective question by reference to what may be reasonably foreseen by a reasonable and prudent employer. It is also crucial that any breach of duty, including breaches of codes of practice or duties imposed by statute, can be shown to be responsible for causing the damage for which compensation is sought.[50]

Judicial review

Public bodies, such as Ofgem, have both statutory duties which must be performed and powers which allow for a large element of discretion. Similarly, the Secretary of State for Business, Energy and Industrial Strategy enjoys wide powers to make decisions about energy policy in the UK. There is usually no right of appeal against a failure to perform a 'power', or as to how that discretion is exercised. However, this does not mean that nothing can be done. Such administrative matters are subject to control by judicial review[51] on the grounds of illegality, irrationality or procedural impropriety (see p239).

The exercise of any 'public' powers by any of the principal bodies discussed in this book and by the Secretary of State for Energy and Climate Change can be subject to judicial review.

An **'illegal decision'** is one where the decision-making body has not been given the legal power to do what it has done – ie, if it has gone outside its remit or what it was set up to do.

An **'unlawful decision'** is where a public body or the Secretary of State for Energy exceed the powers available to them in law and the decision may be challenged in the courts. For example, the Secretary of State was ruled to have no power in law to alter tariffs affecting solar panels retrospectively.[52]

An **'irrational decision'** is one which is so unreasonable that no reasonable authority could make it. In legal jargon, this is 'Wednesbury' unreasonableness, named after the court case in which the principle was established.[53] This principle requires a decision-making body to:

- consider all relevant factors;
- disregard irrelevant factors;
- not act perversely or irrationally.

Challenges may also be brought on grounds of **'procedural impropriety'** where a public body fails to follows its own rules. This can include breaches of 'natural justice', making decisions which are biased or unfair or which have the appearance of unfairness and prejudice to any impartial observer. If a decision-making body fails to adhere to the requirements of natural justice or the 'Wednesbury principles', then its decisions may be challenged in the High Court by way of judicial review. In Scotland, an application is made to the Court of Session. In addition, breaches of the Equality Act 2010 may also be challenged in the High Court.[54]

Breach of one or more of these principles gives the court the power to overturn an authority's decision. It is important to realise that a court cannot overturn a decision simply because it thinks it would have come to a different decision. The court does not put itself in the place of the decision maker, but merely ensures s/he has kept within the boundaries of the law. It is possible for two different, even contradictory, decisions to lie within those boundaries so that it would be equally lawful for the decision maker to choose either. It should also be remembered that judicial review is viewed as a remedy of last resort, and you should normally exhaust all other available remedies before embarking upon it. For example, judicial review may not be commenced if the matter in question is still awaiting the result of an appeal or a decision by Ofgem.[55]

Applying for judicial review is a three-stage procedure.[56] A claimant must serve a letter on the public body or Minister setting out the claim and giving notice of intention to seek judicial review. You must first apply for leave (ie, permission for judicial review) by lodging an application with supporting documents and written evidence. A judge then considers the papers and decides whether leave should be granted – ie, permission to take the case on to a full judicial review hearing. In Scotland, the application is made to the Court of Session.

In England and Wales, the court can make an order overturning a decision ('quashing' order) or requiring the body which is being judicially reviewed to do or not to do something ('mandatory' or 'prohibitory' order) in the same way as an injunction (p236). In Scotland, a decision can be quashed by 'reduction', and a 'declarator' ('declaration' in England and Wales) can be issued establishing the legal position.

It is important to realise that judicial review is a discretionary remedy and that different courts may or may not grant a remedy.

Applications for judicial review in England and Wales must be made promptly to the High Court and, in any event, within three months of the relevant decision. The three-month period can be extended, but only where there are strong mitigating circumstances (delays in the granting of public funding may, on occasion, be such a reason, but this cannot be relied upon). Even if leave is granted, the court may still refuse relief at the full judicial review hearing. In Scotland, applications must be made to the Court of Session; there is no specific time limit, but applications must not be unduly delayed.

Human Rights Act 1998

The Human Rights Act 1998 applies to public authorities whose functions are of a public nature – such as regulators and suppliers. The Public Law Project (see Appendix 1) may be able to comment and advise on the Act. Rights protected under the Human Rights Act are those contained in the European Convention on Human Rights. These include the right to property (Article 1), the right to a fair hearing (Article 6) and the right to privacy and family life (Article 8).

Remission of court fees

If you are on IS, income-based JSA, PC, income-related ESA, UC or working tax credit (but not getting child tax credit), you are entitled to apply for a fee remission when beginning county court proceedings. Proof of entitlement may be established by producing a current benefit letter. If you are turned down on an application for remission of fees, there is normally a right of appeal. An application for a remission is made on Form EX160, which can be downloaded from the HM Courts and Tribunals Service website.

If there is an emergency matter that needs an urgent decision of the court, the court manager can grant a remission without supporting evidence, though you are likely to be required to provide evidence within five days of the remission being given.

6. The Energy Ombudsman

The Energy Ombudsman is an independent body approved by Ofgem. The Ombudsman resolves disputes and complaints after negotiation has failed.

The Ombudsman covers problems related to:
- energy bills;
- sales activity;
- switching gas or electricity supplier;
- the supply of energy to a home, such as power cuts and connections;
- micro generation and feed-in tariffs.

You can only refer your complaint to the Ombudsman if you have tried to resolve it with your energy provider but have received a 'deadlock letter' (see p222) or eight weeks have passed since you first made your complaint to your provider. The Energy Ombudsman expects you to have fully exhausted the supplier's complaints process first.

Most of the complaints to the Ombudsman are about late billing, disputed charges and inaccurate invoices.[57] Ofgem estimates that only 5 per cent of cases that could be referred to the Ombudsman are actually referred.[58]

If you have been unable to resolve your complaint with your energy supplier within eight weeks you may submit your complaint in writing to the Ombudsman. Include all relevant documents. It is also a good idea to include a chronology of events, listing key events in date order. This provides a summary of what happened and when, which assists the Ombudsman in analysing the situation. If you have incurred financial losses, submit copies of receipts, bills and invoices you have had to pay to corroborate what you claim.

If the Ombudsman decides to make an award, and you accept it, your supplier has to abide by the decision. The Ombudsman can ask your supplier to provide any or all of the following:

- a service or some practical action that will benefit you;
- an apology or explanation;
- a financial award up to £10,000 (£10,000 is only payable in exceptional cases; normally awards are much lower).

All of the UK's major energy providers are members of the Ombudsman scheme, which means that they have to abide by any decision that it makes about your complaint.

It is possible to ask for the Ombudsman's decision to be reviewed if you are unhappy with any aspect of it. From 2015, the Ombudsman has also been required to meet the requirements laid down in the Alternative Dispute Resolution for Consumer Disputes (Authorities and Information) Regulation 2015.

For more information, visit www.energy-ombudsman.org.uk.

Notes

1. Remedies available

1 EU Reg 1227/2011 (REMIT)
2 Directive 2009/73/EC concerning common rules for the internal market in gas; and Directive 2009/72/EC concerning common rules for the internal market in electricity.

2. Negotiations

3 Reg 10 GE(CCHS) Regs
4 Ofgem press release, *EDF Energy to pay £3 million following Ofgem investigation into the company's complaints handling arrangements,* 22 August 2014
5 Conditions 27 and 39 SLC
6 *Smith and others v South Eastern Power Networks plc and other cases* [2012] EWHC 2541 QBD(TCC); *Thompson v Smiths Shiprepairers (North Shields) Ltd* [1984] QB 405
7 GE(CCHS) Regs
8 Reg 2 GE(CCHS) Regs
9 Reg 3(1) GE(CCHS) Regs
10 Reg 3(3) GE(CCHS) Regs
11 Ofgem, *Supplier Performance on Consumer Complaints*
12 Reg 4(1) GE(CCHS) Regs
13 Reg 3(3) GE(CCHS) Regs
14 Reg 4(5) GE(CCHS) Regs
15 Reg 4(6) GE(CCHS) Regs
16 Reg 4(4) GE(CCHS) Regs
17 Condition 2 SLC
18 Reg 8(2) EG(S)S Regs
19 s143 EA 2013
20 *R v Director-General of Gas Supply ex parte Smith* [1989] (unreported)

3. Ofgem

21 s144 and Sch 14 EA 2013
22 s27G (1)(b) EA 2013
23 s27H (1)(a)-(c) EA 2013
24 Ofgem, *Financial penalties and consumer redress policy statement,* 6 November 2014
25 Ofgem, *The gas and electricity markets authority's statement of policy with respect to financial penalties and consumer redress under the Gas Act and the Electricity Act,* para 1.4, 31 March 2014

4. Unfair terms and consumer protection

26 s75 and Sch 4 para 34 CRA 2015
27 s62(1) CRA 2015
28 s62(4) CRA 2015
29 s62(5) CRA 2015
30 s68 CRA 2015
31 Sch 5 CRA 2015
32 Sch 5 para 4 CRA 2015
33 Sch 2 CRA 2015
34 Reg 2(1) CPUT Regs
35 Reg 5 CPUT Regs
36 *R(on the application of Surrey Trading Standards) v Scottish and Southern Energy plc* [2012] CA, Criminal Division, 16 February, 16 March 2012
37 Art 4 and Sch 1 Public Bodies (Abolition of the National Consumer Council and Transfer of the Office of Fair Trading's Functions in Relation to Estate Agents etc) Order 2014 No.631, amending GA 1986 and EA Act 1989
38 CPR Part 30.3
39 *McKenzie v McKenzie* [1970] 3 WLR 472
40 *Oak Cash and Carry Ltd v British Gas Trading Ltd* [2016] EWCA Civ 153
41 *Mitchell v News Group Newspapers Ltd* [2013] EWCA Civ 1537
42 *Ferguson v British Gas* [2009] EWCA Civ 46
43 *Say v British Gas Services Ltd* [2011] All ER (D) 216
44 *Gwenter v Eastern Electricity plc* [1994] *Legal Action,* August 1995, p19
45 HCTS, Court Fees for the High Court, county court and family court, 2 August 2016
46 The principles are set out in *American Cyanamid v Ethicon Ltd* [1975] AC 396
47 *Gwenter v Eastern Electricity plc* [1994] *Legal Action,* August 1995, p19
48 *Faulkner v Yorkshire Electricity plc* [1994] *Legal Action,* February 1995, p23; *Gwenter v Eastern Electricity plc* [1994] *Legal Action,* August 1995, p19
49 See *Red Star Pub Company (WRII) Ltd and Others v Scottish Power Ltd* [2016] CSOH 100
50 *Smith and others v South Eastern Power Networks plc and other cases* [2012] EWHC 2541 QBD(TCC)

51 Part 54 CPR
52 *Secretary of State for Energy and Climate Change v Friends of the Earth and others* [2012] EWCA Civ 28
53 *Associated Provincial Picture Houses v Wednesbury Corporation* [1948] 1 KB 223
54 Ofgem, *Consumer Vulnerability Strategy,* 4 July 2013
55 *R (on the application of Summerleaze Ltd) v Secretary of State for Energy and Climate Change* [2015] EWHC 1729 (Admin)
56 Part 54 CPR

6. The Energy Ombudsman

57 Ombudsman data sheets, 2016 quarter 2
58 Complaints to energy companies research 2014, available at www.ofgem.gov.uk/sites/default/files/docs/2014/09/ofgem_complaints_report_final_8_august_2014.pdf

Appendix 1

Useful addresses and publications

Energy suppliers' contact numbers

British Gas	0800 048 0202
Co-operative Energy	0800 954 0693
Ebico	0800 458 7689
Ecotricity	0345 555 7100
EDF Energy	0800 096 9000
(including customers of Seeboard and SWEB)	
E.ON	0843 506 9877
First Utility	01926 320 700
Good Energy	0800 254 0000
Loco2energy	0845 074 3601
npower	0800 073 3000
OVO Energy	0800 599 9440
Scottish Hydro	0345 026 0655
Scottish Power	0800 027 0072
Spark Energy	0345 034 7474
SSE	0345 026 2658
(including customers of Equigas/Equipower, Atlantic, Severn Trent Energy, Southern Electric)	
SWALEC	0345 026 0655
Utilita	0345 207 2000
Utility Warehouse	0333 777 0777

Fuel campaigning and information organisations

Energy Saving Trust (EST)
Helpline: 0300 123 1234
www.energysavingtrust.org.uk

EST England
21 Dartmouth Street
London SW1H 9BP
Tel: 020 7222 0101

EST Scotland
2nd Floor, Ocean Point 1
94 Ocean Drive
Edinburgh EH6 6JH
Tel: 0808 808 2282

EST Wales
1 Caspian Point, Caspian Way
Cardiff Bay CF10 4DQ
Tel: 0800 512 012

National Energy Action
Level 6, West One
Forth Banks
Newcastle upon Tyne NE1 3PA
Tel: 0191 261 5677
www.nea.org.uk

Energy Action Scotland
Suite 4A Ingram House
227 Ingram Street
Glasgow G1 1DA
Tel: 0141 226 3064
www.eas.org.uk

National Energy Foundation
Davy Avenue
Knowlhill
Milton Keynes MK5 8NG
Tel: 01908 665 555
www.nef.org.uk

Fuel and energy industry bodies

Ofgem
9 Millbank
London SW1P 3GE
Tel: 020 7901 7295
www.ofgem.gov.uk

Ofgem – Scotland
3rd Floor, Cornerstone
107 West Regent Street
Glasgow G2 2BA
Tel: 0141 331 2678

Ofgem – Wales
1 Caspian Point, Caspian Way
Cardiff Bay CF10 4DQ

Gas Safe Register
PO BOX 6804
Basingstoke RG24 4NB
Tel: 0800 408 5500
www.gassaferegister.co.uk

Energy UK
Charles House
5-11 Regent Street
London SW1Y 4LR
Tel: 020 7930 9390
www.energy-uk.org.uk

Solid Fuel Association
7 Swanwick Court
Alfreton DE55 7AS
Helpline: 01773 835400
www.solidfuel.co.uk

Consumer and debt advice and information

Citizens Advice consumer service
Post Point 24
Town Hall
Walliscote Grove Road
Weston super Mare
North Somerset
BS23 1UJ
Helpline: 03454 04 05 06
Welsh speaking line: 03454 04 05 05
www.adviceguide.org.uk

Competition and Markets Authority
Victoria House
37 Southampton Row
London WC1B 4AD
Tel: 020 3738 6000
www.gov.uk/government/organisat-
ions/competition-and-markets-au-
thority

Which?
www.which.co.uk

Legal advice

Civil Legal Advice
Helpline: 0345 345 4345
www.gov.uk/civil-legal-advice

Law Centres Federation
www.lawcentres.org.uk

Public Law Project
150 Caledonian Road
London N1 9RD
Helpline: 020 7843 1260
www.publiclawproject.org.uk

Specialist advice organisations

Age UK
Tavis House
1-6 Tavistock Square
London WC1H 9NA
Helpline: 0800 169 6565
www.ageuk.org.uk

Child Poverty Action Group
30 Micawber Street
London N1 7TB
Tel: 020 7837 7979
www.cpag.org.uk

Shelter
88 Old Street
London EC1V 9HU
Helpline: 0808 800 4444
www.shelter.org.uk

Turn2us
Helpline: 0808 802 2000
www.turn2us.org.uk

Useful publications

The following books are available from CPAG, tel: 020 7812 5227; email: bookorders@cpag.org.uk. An order form, as well as details of CPAG's other books and subscription services, and details of the prices for members and CABx, is at www.shop.cpag.org.uk.

Welfare Benefits and Tax Credits Handbook (CPAG)
£61/£15 for claimants (2016/17, April 2016)

Welfare Benefits and Tax Credits Handbook Online (CPAG)
Includes the full text of the *Welfare Benefits and Tax Credits Handbook*

updated throughout the year. Annual subscription £70 per user (bulk discounts available). More information at www.shop.cpag.org.uk.

Debt Advice Handbook (CPAG)
£26 (11th edition, July 2015)

Council Tax Handbook (CPAG)
£22 (11th edition, February 2016)

Benefits for Migrants Handbook (CPAG)
£36 (8th edition, October 2016)

Universal Credit: what you need to know (CPAG)
£15 (3rd edition, autumn 2015)

Winning Your Benefit Appeal: what you need to know (CPAG)
£15 (2nd edition, autumn 2016)

Personal Independence allowance: what you need to know (CPAG)
£15 (2nd edition, July 2016)

For online subscriptions and orders up to £10 in value; for order value £10.01–£100, add a flat rate charge of £3.99; for order value £100.01–£400, add £7.49; for order value £400+, add £11.49.

Appendix 2

Reading your meter

Electricity

There are four types of electricity meter in common use.

Standard credit meter
A standard meter measures electricity consumption in kilowatt hours (kWh) – the number of units of energy used in an hour. With this type of meter, all electricity units are charged at the same rate 24 hours a day. Most standard meters have an electronic or digital display showing a row of numbers. Older meters may have a dial display with four or more dials, each with a pointer.

Variable rate credit meter
A variable rate meter operates on the same principle as a standard meter but gives more than one reading display – ie, to show daytime, normal or peak electricity use, overnight or low off-peak use and (if appropriate) controlled circuit use. Customers with a variable rate credit meter will have either one or two meters showing up to three sets of numbers. The majority of these meters will have electronic or digital displays showing rows of numbers. A few customers may still have two dial display meters installed – one each for peak and off-peak consumption.

Prepayment meter
A prepayment meter measures electricity use in exactly the same way as a credit meter. A prepayment meter has a digital display screen which can show a range of information.

Smart meter
A smart meter is an electronic meter. It operates in the same way as a credit meter in terms of registering electricity consumption. Many smart meters have visual displays to highlight energy consumption levels. Most have a digital display which can show a range of information.

Reading your meter

The numbers on electronic and digital displays should always be read from left to right. Write down the first five numbers shown. Red numbers, or numbers after a decimal point, should be ignored.

If you want to work out how much electricity you use, write down the numbers and take a note of the date. The next time you take a reading subtract the second reading from the first and you will know how many units (kilowatt hours) you have used in the period since you took your first reading.

The same principle applies for reading dial meters. These should also be read from left to right, ignoring the final (usually red) dial. Write down the number closest to each pointer. If the pointer is between two numbers, write down the lower number, but if the pointer is between 9 and 0, write down 9.

If there are two rows of numbers, the top row is usually for off-peak and may be marked 'low' or 'night'. The bottom row is usually for peak and may be marked 'normal' or 'standard'.

Some variable rate meters have only one digital display. This type of meter will usually show the charging rate that's currently in use. These meters should have a button that will cycle through the readings for the different rates.

If you are working out how much electricity you have used, make sure you note clearly which reading is which.

Prepayment meters normally display the amount of credit remaining for use. To obtain a reading from a prepayment meter, you will have to press a button on the meter to change the digital display. Pressing the button repeatedly will allow you to cycle through the display screens (to return to the original screen, stop pressing the button). Every prepayment meter provides a range of information but all are configured slightly differently. However, most use letters to count the display screens and include displays for:

- current credit;
- total credit accepted – ie, amount topped up onto meter;
- reading for rate 1;
- price per unit for rate 1;
- reading for rate 2 (if appropriate);
- price per unit for rate 2;
- standing charge;
- amount available for emergency credit;
- debt repayment level (if appropriate).

Note: you may have to insert your key/card/token to view all the displays.

Gas

There are three types of gas meter in common use.

Standard credit meter

The majority of gas customers have a credit meter which records the amount of gas used. Gas consumption is measured in units. For many older meters – imperial meters – gas usage is measured in cubic feet. For newer metric meters, gas usage is measured in cubic metres.

Most standard meters have an electronic or digital display showing a row of four or five numbers. Older meters may have a dial display with four or more dials, each with a pointer.

Prepayment meter

A prepayment meter measures gas use in exactly the same way as a credit meter. A prepayment meter has a digital display screen which can show a range of information.

Smart meter

A smart meter is an electronic meter. It operates in the same way as a credit meter in terms of registering gas consumption. Many smart meters have visual displays to highlight energy consumption levels. Most have a digital display which can show a range of information.

Reading your meter

The numbers on electronic and digital displays should always be read from left to right. Red numbers, or numbers after a decimal point, should be ignored.

If you want to work out how much gas you use, write down the numbers and take a note of the date. The next time you take a reading, subtract the second reading from the first and you will know how many units you have used in the period since you took your first reading.

The same principle applies for reading dial meters. These should also be read from left to right, ignoring the final (usually red) dial. Write down the number closest to each pointer. If the pointer is between two numbers, write down the lower number, but if the pointer is between 9 and 0, write down 9.

Prepayment meters normally display the amount of credit remaining for use. To obtain a reading from a prepayment meter, you will have to press a button (this may be marked 'A') on the meter to change the digital display. Pressing the button repeatedly will allow you to cycle through the display screens. Every prepayment meter provides a range of information but all are configured slightly differently. However, most use letters to count the display screens and include displays for:

- current credit;
- last credit (most recent amount topped up. Displays on some meters may also show how much was paid towards gas consumption, emergency credit repayment and debt repayment);
- total credit accepted – ie, amount topped up onto meter;

- reading;
- price per unit;
- standing charge;
- amount available for emergency credit;
- debt repayment level (if appropriate);
- debt remaining.

Note: you may have to insert your key/card/token to view all the displays.

Submitting meter readings

If you want to provide meter readings to your supplier, you can do this online or over the phone. This will help ensure that any bills you receive are accurate, that weekly/monthly payment amounts are appropriate and it will help prevent debt building up on your account.

Calculating your costs

Before you can calculate the cost of your electricity and gas consumption, you will need to know what your tariff (the amount you pay for every kilowatt hour) is. You will find the specific name for your tariff on your fuel bill or your annual statement. Alternatively, you can phone your supplier to ask. Depending on the type of tariff you have, you will usually have a standing charge to pay along with the cost of your ongoing fuel use. This will also be shown on your bill and annual statement as a daily charge.

For gas consumption, you need to check your meter to see whether you have an old imperial meter or a newer metric meter. If it is an imperial meter measuring gas in cubic feet, it will usually have the words 'cubic feet' or 'Ft³' shown somewhere on the front of the meter. If it is a metric meter measuring gas in cubic metres it will usually show the words 'cubic metres' or 'M³'.

Your tariff for gas will be in kilowatt hours (kWh), so the readings from your gas meter need to be converted into kWh, so that you can then work out how much the fuel you use is costing. You can do this by:

- multiplying units used by 2.83 to give the number of cubic metres of gas used (if the meter is a newer metric one measuring gas in cubic metres this part of the calculation is not needed);
- multiplying by the temperature and pressure figure (1.02264);
- multiplying by calorific value (approximately 39.5, though the exact calorific value can be found on a gas bill);
- dividing by 3.6 to get the number of kWh.

Appendix 3

. . . .

Understanding fuel bills and annual statements

. . . .

Example fuel bills and annual statements

Each supplier's fuel bills will look slightly different, but the amount and type of information that must be included is standard. uSwitch and Moneysupermarket have a sample of electricity, gas and dual fuel bills from some of the biggest suppliers at www.uswitch.com/gas-electricity/guides/energy-bills/ and www.moneysupermarket.com/gas-and-electricity/hubs/energy-bills-explained/

Ofgem has provided suppliers with a standard template for annual statements, so that these are all produced in the same format. The template is contained within the Standard Licence Conditions.

. . . .

Suppliers are required to provide you with specific information on your electricity and gas bills and annual statements. This is part of Ofgem's reforms to establish a 'simpler, clearer, fairer' market for consumers.

If you are a credit meter customer, you should receive a bill every quarter (available online if you have opted for paperless billing). The information that must appear on your bill includes:

- specific tariff name;
- whether your reading/bill is actual or estimated;
- details of your energy use;
- details of how much your energy costs, including unit cost and standing charges applied;
- an explanation of how your bill has been calculated;
- usage comparison between the current period and the same period a year ago.

Your bill must also include information advising whether your supplier has a cheaper alternative tariff, a personal projection, a tariff information label (TIL), a tariff comparison rate (TCR) and your meter reference number.

Personal projection

The aim of a personal projection is to indicate what your energy bills will cost over the coming year and to show whether you are on your supplier's cheapest tariff. If this is not the case, the information is printed on your bill along with the cheapest alternative. You can also use your personal projection to see if you can find a cheaper tariff with a new energy provider, but note that for a realistic and accurate comparison, you should obtain your actual use in kilowatt hours (kWh) from your annual statement. If you don't have this to hand, your supplier can provide this information.

Tariff information label

A TIL lists the key facts about your tariff, to enable you to compare it with others. The summarised version on a fuel bill includes:
* tariff name;
* payment method;
* end date;
* exit fees, if applicable (the fees payable if you terminate your tariff before the official end date);
* assumed annual use.

Your annual statement also includes:
* supplier name;
* tariff type (eg, fixed, variable);
* payment method;
* unit rate (price of energy per kWh);
* standing charge, if applicable (the per day fixed cost associated with providing your energy supply, such as meter readings and maintenance);
* price guarantee (date when the price you're currently paying is guaranteed until);
* additional products and services (eg, loyalty points).

As with the personal projection, if you want an accurate comparison, get your actual use in kWh from your annual statement. If you do not have this to hand, your supplier can provide this information.

Tariff comparison rate

The TCR is a figure that can be used to compare the cost of the tariff you are on against alternative tariffs from your own or other suppliers. The TCR is not based on your personal consumption and should be used as a guide only. It is a way of breaking down an electricity or gas tariff to a single unit rate, shown as p/kWh, to allow you to view different tariff rates side by side.

Calculating the tariff comparison rate

1. Calculate the supplier's standing charge as an annual amount in pence.
2. Take the supplier's energy unit rate in p/kWh and multiply it by 3,200 kWh for electricity and 13,500 kWh for gas (Ofgem's average domestic consumption figures) and add it to the standing charge figure in step 1.
3. Subtract any discounts on the tariff.
4. Add VAT at 5 per cent (if not already included in the unit rate and standing charge).
5. Divide this calculated figure by 3,200 or 13,500 for electricity and gas respectively.

The difficulty with the TCR is there may be some occasions when one tariff shows a lower TCR rate than another supplier's tariff, but if you are a very low energy user it may have been cheaper to take the tariff with the higher TCR if the standing charge on that tariff is lower, for example. Not everyone has access to the internet or has their energy consumption to hand, so the TCR may provide a useful starting point, but you should not rely on the TCR as the sole way to compare energy prices.

Please note that the TCR is not yet available for consumers with Economy 7, Economy 10 or other time-of-use meters.

Meter reference numbers

Every gas and electricity meter has a unique reference number that effectively links the meter to a specific address. For electricity meters, this is known as the meter point administration number (MPAN). The format of a MPAN is standard and consists of 21 digits. For gas it is known as the meter point reference number (MPRN). The format of a MPRN is also fairly standard and consists of between six and 10 digits. Occasionally, these are also referred to simply as 'meter numbers' or as 'M' or 'S' (supply) numbers. The reference number(s) must be shown on your fuel bills and your annual statement. If you are going to switch supplier, your new supplier will need your meter reference number(s). If you cannot find your MPAN number, contact your district network operator. For your MPRN number, phone the Meter Point Reference Helpline on 0870 608 1524.

Appendix 4

National standards for enforcement agents

Taking Control of Goods: National Standards

Issued 6 April 2014 by the Ministry of Justice.

Vulnerable situations

70. Enforcement agents/agencies and creditors must recognise that they each have a role in ensuring that the vulnerable and socially excluded are protected and that the recovery process includes procedures agreed between the agent/agency and creditor about how such situations should be dealt with. The appropriate use of discretion is essential in every case and no amount of guidance could cover every situation. Therefore the agent has a duty to contact the creditor and report the circumstances in situations where there is evidence of a potential cause for concern.

71. If necessary, the enforcement agent will advise the creditor if further action is appropriate. The exercise of appropriate discretion is needed, not only to protect the debtor, but also the enforcement agent who should avoid taking action which could lead to accusations of inappropriate behaviour.

72. Enforcement agents must withdraw from domestic premises if the only person present is, or appears to be, under the age of 16 or is deemed to be vulnerable by the enforcement agent; they can ask when the debtor will be home – if appropriate.

73. Enforcement agents must withdraw without making enquiries if the only persons present are children who appear to be under the age of 12.

74. A debtor may be considered vulnerable if, for reasons of age, health or disability they are unable to safeguard their personal welfare or the personal welfare of other members of the household.

75. The enforcement agent must be sure that the debtor or the person to whom they are entering into a controlled goods agreement understands the agreement and the consequences if the agreement is not complied with.

76. Enforcement agents should be aware that vulnerability may not be immediately obvious.

77. Some groups who might be vulnerable are listed below. However, this list is not exhaustive. Care should be taken to assess each situation on a case by case basis.

- the elderly;
- people with a disability;
- the seriously ill;
- the recently bereaved;
- single parent families;
- pregnant women;
- unemployed people; and,
- those who have obvious difficulty in understanding, speaking or reading English.

78. Wherever possible, enforcement agents should have arrangements in place for rapidly accessing interpretation services (including British Sign Language), when these are needed, and provide on request information in large print or in Braille for debtors with impaired sight.

Appendix 5

Qualifying energy improvements

The following energy efficiency improvements are permitted by the Schedule to the Green Deal (Qualifying Energy Improvements) Order 2012:

- air source heat pumps
- biomass boilers
- biomass room heaters (with radiators)
- cavity wall insulation
- chillers
- cylinder thermostats
- draught proofing
- duct insulation
- external wall insulation systems
- fan-assisted storage heaters
- flue gas heat recovery devices
- gas-fired condensing boilers
- ground source heat pumps
- heating controls for wet central heating systems or warm air systems
- heating ventilation and air-conditioning controls (including zoning controls)
- high performance external doors
- hot water controls (including timers and temperature controls)
- hot water cylinder insulation
- hot water showers
- hot water systems
- hot water taps
- internal wall insulation systems (for external walls)
- lighting systems, fittings and controls (including rooflights, lamps and luminaires)
- loft or rafter insulation (including loft hatch insulation)
- mechanical ventilation with heat recovery systems
- micro combined heat and power
- micro wind generation
- oil-fired condensing boilers
- photovoltaics
- pipework insulation
- radiant heating

replacement glazing
roof insulation
room in roof insulation
sealing improvements (including duct sealing)
secondary glazing
solar blinds, shutters and shading devices
solar water heating

transpired solar collectors
under-floor heating
under-floor insulation
variable speed drives for fans and pumps
warm-air units
waste water heat recovery devices attached to showers
water source heat pumps

Appendix 6

....

Draft court claim

This appendix gives a precedent for a court claim in England and Wales against an electricity supplier for breach of its duty to supply or against a gas or electricity supplier for breach of contract. Hopefully, this is useful for advisers who are not familiar with this area of the law. This draft claim is put in the county court which is where most claims will be heard.

IN THE _____ COUNTY COURT

Case No _____

BETWEEN:

A.N. OTHER   Claimant

and

-X- ELECTRICITY PLC
or
-Y- GAS PLC   Defendant

PARTICULARS OF CLAIM

1. The Defendant is a [*gas/electricity*] supplier and is licensed to supply [*gas/electricity*] to an authorised area, within the meaning of the [*Gas Act 1986/Electricity Act 1989*]. The said authorised area includes the premises known as and situated at [*your address*] ('the premises') which are [*owned/occupied*] by the Claimant.

2. [*For electricity supplied under contract:*]
The Defendant supplies electricity to the Claimant in accordance with its Designated Supply Contract as defined by Condition 42 of the Second Tier Electricity Supply Licence.

[*For gas:*]
The Claimant has been a customer of the Defendant since [*insert date when you started paying the bill at present address*] and is now supplied pursuant to a contract in accordance with the Gas Acts 1986 and 1995.

3. [*For gas or electricity supplied under contract:*]
It is an express term of the said contract that the Defendant shall give a supply and continue to give a supply to the Claimant.

4. In breach of the [*provisions of the Electricity Act 1989/said term of the contract*] set out in paragraph 3 above, the Defendant has failed to [*give/continue supply to the premises*].

PARTICULARS OF BREACH
[*Set out here concisely the facts of the situation on which you would rely as supporting your case in court – below is an example*]

On 31 October 2016, representatives of the Defendant came to the premises and found a hole in the side of the [*electricity/gas*] meter. The meter was removed by the said representatives on the same day. The Claimant has been without a supply since then.

The Defendant demanded the sum of £ [INSERT SUM] for the damage to the meter and for disconnection and reconnection charges and this was paid on 11 November 2016.

The Defendant's claim, by letter dated 15 November 2016, that the Claimant owes the further sum of £[INSERT SUM] in respect of [*electricity/gas*] supplied but not registered on the damaged meter and refuse to reconnect supply until this sum is paid. The Claimant does not know how this sum is calculated and genuinely disputes that any part of it is owed.

5. Further, the Defendant is in breach of [*Schedule 6 paragraph 1(9) of the Electricity Act 1989/Schedule 2B paragraph 7(5) of the Gas Act 1986 – cannot disconnect when sum genuinely in dispute*].

PARTICULARS OF BREACH
The Claimant relies on the particulars set out in paragraph 4 above.

6. By reason of the matters aforesaid, the Claimant has suffered loss, damage, nuisance, inconvenience, anxiety and distress.

PARTICULARS OF DAMAGE
[*Again, what follows below is an example*]

The Claimant lives at the premises with her husband, John, and two daughters: Helen (5 years old) and Joanna (3 years old). John and Joanna both have asthma which is made worse by cold conditions.

The Claimant has been unable to use the central heating system at the premises and has had to buy coal and paraffin to heat the premises this costs an average of £00.00 a [*day/week*].

Because there is no working cooker or fridge/freezer, the Claimant and her family have had to eat meals out. On average, £00.00 more is spent on each meal than if it had been made at home.

7. Further, the Claimant claims interest pursuant to section 69 of the County Courts Act 1984 on such sums as may be found due to the Claimant, at such rate and for such period as the court shall think fit.

AND THE CLAIMANT CLAIMS:
1. A declaration that the Claimant does not owe the Defendant the sum of £00.00 or any other amount in respect of the supply of [*electricity/gas*].
2. An injunction requiring the Defendant to install a new meter at the premises and to restore supply forthwith.
3. Damages.
4. Further and other relief as the Court sees fit.
5. Costs.
6. Interest pursuant to the County Courts Act 1984 section 69 as aforesaid.

Dated this _____ day of _____ 201 ____

Signed _____

Solicitor for the Claimant

Appendix 7

Abbreviations used in the notes

AAC	Administrative Appeals Chamber
AC	Appeal Cases
All ER	All England Reports
Art(s)	Article(s)
CA	Court of Appeal
CPR	Civil Procedure Rules
Crim LR	Criminal Law Reports
EWCA Civ	England and Wales Court of Appeal (Civil Division)
EWHC	England and Wales High Court
LC	Lands Chamber
Para	paragraph
QB	Queen's Bench Reports
r(r)	rule(s)
Reg(s)	Regulation(s)
s(s)	section(s)
Sch(s)	Schedule(s)
ScotCS	Scottish Court of Session
SC	Supreme Court
SLC	Standard Licence Conditions 2014
UKSC	United Kingdom Supreme Court
UKUT	United Kingdom Upper Tribunal
WLR	Weekly Law Reports

Acts of Parliament

CEARA 2007	The Consumers, Estate Agents and Redress Act 2007
CRA 2015	Consumer Rights Act 2015
DA 2015	Deregulation Act 2015
DPA 1972	The Defective Premises Act 1972
EA 1989	The Electricity Act 1989

EA 2010	The Electricity Act 2010
EA 2013	Energy Act 2013
GA 1986	The Gas Act 1986
GA 1995	The Gas Act 1995
HA 1980	The Housing Act 1980
HA 1985	The Housing Act 1985
HA 1996	The Housing Act 1996
H(S)A 2006	The Housing (Scotland) Act 2006
LA 2011	The Localism Act 2011
LG(MP)A 1976	The Local Government (Miscellaneous Provisions) Act 1976
LTA 1985	The Landlord and Tenant Act 1985
RA 1977	The Rent Act 1977
RE(GEB)A 1954	Rights of Entry (Gas and Electricity Boards) Act 1954
R(S)A 1984	The Rent (Scotland) Act 1984
TA 1968	Theft Act 1968
UA 2000	The Utilities Act 2000

Regulations and other statutory instruments

ASTNPR(E) Regs	The Assured Shorthold Tenancy Notices and Prescribed Requirements (England) Regulations 2015 No.1646
CP(CCCBP) Regs	The Consumer Protection (Cancellation of Contracts Concluded Away from Business Premises) Regulations 1987 No.2117
CPUT Regs	The Consumer Protection From Unfair Trading Regulations 2008 No.1277
E(CSP) Regs	The Electricity (Connection Standards of Performance) Regulations 2015 No.698
E(PM) Regs	The Electricity (Prepayment Meter) Regulations 2006 No.2010
E(SP) Regs	The Electricity (Standards of Performance) Regulations 2015 No.699
EE(PRP)(EW) Regs	The Energy Efficiency (Private Rented Property) (England and Wales) Regulations 2015 No.962
EG(B) Regs	The Electricity and Gas (Billing) Regulations 2014 No.1648
EG(SP)S Regs	The Electricity and Gas (Standards of Performance) (Suppliers) Regulations 2015 No.1544

● ●

EPB(CI)(EW) Regs	The Energy Performance of Buildings (Certificates and Inspections) (England and Wales) Regulations 2007 No.991
EPB(EW) Regs	The Energy Performance of Buildings (England and Wales) Regulations 2012 No.3118
ES Regs	The Electricity Supply Regulations 1988 No.1057
ESQC Regs	The Electricity Safety, Quality and Continuity Regulations 2002 No.2665
FP(E) Regs	The Fuel Poverty (England) Regulations 2014 No. 3220
G(PM) Regs	The Gas (Prepayment Meter) Regulations 2006 No.2011
G(SP) Regs	The Gas (Standards of Performance) Regulations 2005 No.1135
GE(CCHS) Regs	The Gas and Electricity (Consumer Complaints Handling Standards) Regulations 2008 No.1898
GS(IU) Regs	The Gas Safety (Installation and Use) Regulations 1998 No.2451
GS(RE) Regs	The Gas Safety (Rights of Entry) Regulations 1983 No.1575
HEAS(S) Regs	The Home Energy Assistance Scheme (Scotland) Regulations 2013 No.148
HB Regs	The Housing Benefit Regulations 2006 No.213
HB(SPC) Regs	The Housing Benefit (Persons who have attained the qualifying age for State Pension Credit) Regulations 2006 No.214
RR(HA)(EW)O	The Regulatory Reform (Housing Assistance) (England and Wales) Order 2002 No.1860
SFCWP Regs	The Social Fund Cold Weather Payments (General) Regulations 1988 No.1724
SFWFP Regs	The Social Fund Winter Fuel Payments Regulations 2000 No.729
SS(C&P) Regs	The Social Security (Claims and Payments) Regulations 1987 No.1968
UC Regs	The Universal Credit Regulations 2013 No.376
UC,PIP,JSA&ESA (CP) Regs	The Universal Credit, Personal Independence Payment, Jobseeker's Allowance and Employment and Support Allowance (Claims and Payments) Regulations 2013 No.380
WHD(MA) Regs 2016	The Warm Home Discount (Miscellaneous Amendments) Regulations 2016 No.806

Index

Index